Optoelectronic Switching Systems in Telecommunications and Computers

ELECTRO-OPTICS SERIES

Series Editor: Dr. Herbert A. Elion
Chief Executive Officer, International Communications and Energy, Inc.
and
President, AETNA Telecommunications Consultants
Centerville, Massachusetts

Volume 1   Molecular Electro-Optics: Part 1—Theory and Methods; Part 2—Applications to Biopolymers, *edited by Chester T. O'Konski*

Volume 2   Fiber Optics in Communications Systems, *by Glenn R. Elion and Herbert A. Elion*

Volume 3   Electro-Optics Handbook, *by Glenn R. Elion and Herbert A. Elion*

Volume 4   Optoelectronic Switching Systems in Telecommunications and Computers, *by Herbert A. Elion and V. N. Morozov*

Other volumes in preparation

Optoelectronic Switching Systems in Telecommunications and Computers

HERBERT A. ELION
International Communications and Energy, Inc.
Aetna Telecommunications Consultants
Centerville, Massachusetts

V. N. MOROZOV
P. N. Lebedev Physical Institute
Moscow
Union of Soviet Socialist Republics

CRC Press
Taylor & Francis Group
Boca Raton London New York

CRC Press is an imprint of the
Taylor & Francis Group, an **informa** business

First published 1984 by Marcel Dekker, Inc.

Published 2019 by CRC Press
Taylor & Francis Group
6000 Broken Sound Parkway NW, Suite 300
Boca Raton, FL 33487-2742

© 1984 by Taylor & Francis Group, LLC
CRC Press is an imprint of Taylor & Francis Group, an Informa business

First issued in paperback 2019

No claim to original U.S. Government works

ISBN-13: 978-0-367-45184-4 (pbk)
ISBN-13: 978-0-8247-7163-8 (hbk)

Visit the Taylor & Francis Web site at
http://www.taylorandfrancis.com

and the CRC Press Web site at
http://www.crcpress.com

Library of Congress Cataloging in Publication Data

Elion, Herbert A., [date].
  Optoelectronic switching systems in telecommunications
and computers.

    (Electro-optics series ; v. 4)
    Includes index.
    1. Telecommunication--Switching systems. .
2. Optoelectronic devices, I. Morozov, V. N. (Valentin
N.), [date]. II. Title. III. Series.
TK5102.5.E44   1984   621.38'0414   84-1779
ISBN 0-8247-7163-X

Preface

The purpose of this book is to present the general engineering
considerations that have resulted in a fundamental change in telecommunications computer networks. The term for such networks
is currently C^4I: computers, communications, command, control,
and intelligence. Typical types of information are telephone, data,
facsimile, and video. Such communications of intelligence, sometimes at an appreciable distance, involve many disciplines that
work together to form a system. Prior to 1980, telecommunications were traditionally broken down into two major engineering
categories, namely, transmission and switching (or three if terminals are included).

The advent of data communications and especially digital telephony led to the present indistinct separation of disciplines. In
some cases, as in integrated digital telephone networks and packet-
switched data networks, the change was revolutionary, and the
dividing lines essentially disappeared. Not only is a change occurring in processing information but in the intrinsic nature of
the hardware, which is not just electronic but optoelectronic and
photonic.

This book concentrates on the switching aspects. In telecommunications, such switching is the function of establishing and
releasing connections among transmission paths. To provide transmission paths, the originating transmission goes from a terminal to
a switching center, then to switching centers, and to the eventual
recipient.

For a telecommunications system to satisfy a fluctuating demand
that can be predicted with at least a limited degree of accuracy,
the service requires high performance standards. Since this may
be expensive, the optimization of the network and equipment performance is of considerable importance.

As the reader will note, the optoelectronic processors discussed could readily be, or become part of, a telecommunications system or a computer system. Indeed, fusion into both, and into C^4I systems, is highly likely before the end of the century on a pervasive basis, rather than the more occasional basis of today.

In the present book, optoelectronic switching is emphasized in the fusion into traditional telecommunications of the present. The "wedded" technologies are treated in their revolutionary aspect as they affect network design and hardware. With the use of parallel and series optical operations, logic and computational speed, economics, and size are undergoing substantial improvements. Optoelectronic matrix switching, for example, shows promise of going to large dimensions both in heterojunction and homojunction switching photodiode forms, with isolations of better than 80 dB and frequencies well beyond 1 GHz.

Even submicrometer lithography processes show promise of reaching 0.1 μm from the present 0.5- to 2-μm sizes or substrate semiconductor chips. But the further significant parallel processing advantages arise from the photonics nature of the processing.

Services may be distributed or switched, and central or customer switching may be used. Multiplexing may be electronic, or optical, or some optimal combination of both. Links may be bidirectional or individual.

We are indeed at a point where 1000 gigaoperations per second are possible in the near future with photonics and optoelectronics. This text gives some of the general engineering considerations of such improvements.

The authors gratefully acknowledge the contribution of Academician N. Basov in introducing the basic subject matter and giving a perspective on the importance of optoelectronics and photonics for the handling of information.

Herbert Elion thanks Kathleen and Jonathan Elion of ECG in Lexington, Kentucky for deciphering and correcting his original manuscript as well as clarifying the text. He also thanks Dr. Glenn R. Elion for some material upon which they have worked together over the past several years. Sheila Thall contributed fine technical editing, limited only by the material supplied. Dr. Edouard Y. Rocher assisted with his expertise in local area networks.

Finally, both authors gratefully acknowledge the perseverance on the part of all concerned to bring this book to publication.

Herbert A. Elion
V. N. Morozov

Introduction

Computer engineering is one of the most important factors accelerating social, scientific, and industrial development. Numerous applications may be cited where computers have opened the only possible way to solution: involved scientific and engineering calculations, spacecraft control, economic problems, scientific and engineering data systems, and control of scientific experiments. The area of computer applications is still growing.

The progress of electronic engineering is impressive, and it is quite reasonable to ask why new hardware should be developed. Is it not possible to design computers relying on the existing devices?

The answer is as follows. Computer throughput largely depends on the speed of arithmetic and logic circuits. On the average, each decade brings improvement of computer performance by the order of 3, beginning from 0.3 sec in 1944. Is it plausible to continue such a thousandfold improvement and to design a computer having an addition time of 0.3 nsec? Many experts believe that with conventional electronic elements this is unlikely because of limited signal speed in interconnections. The speed of arithmetic and logical devices grows essentially slower than that of the elements and is about $\tau^{0.6}$, where τ is element speed. This may be understood if one takes into account that the total element response delay in a logical path is only a fraction (usually from 0.3 to 0.5) of the circuit delay, the balance being for interconnection delays. As interconnection length is proportional to $N^{0.25}$, where N is the number of gates in a logic integrated circuit, reduction of interconnections by 5 times requires an increase of element integration by 600 times. Understandably, the advent of very fast VLSI (very large scale integrated) circuits where a reduction of minimal topological dimensions from 2.5 to 0.5 μm results in an increase in the gate count of 5 X 10^3 to 2.5 X 10^5 per circuit does not solve the problem radically because gate delay drops from 25-5 nsec only to 5-1 nsec.

Thus, an essential increase in computer performance may be achieved through design of structures enabling concurrent execution of multiple computations, i.e., parallel data processing.

For a long time the idea of attaining high speed through concurrent execution of many binary operations has been the object of both theoretical and experimental studies in electronic computer engineering, which may be exemplified by such systems as Illiac-4 and Cray-1.

It should be noted that computer structures are closely related to the hardware which is planned to be used in designs. Therefore, any attempt to use old hardware for future computers or to repeat old structures with new hardware handicaps further progress in computer engineering.

This suggests that new approaches to computations require the development of new hardware. Naturally, the problem arises concerning the hardware for future high-throughput parallel systems. There is, evidently, only one answer: Many difficulties characteristic of electronics may be avoided by using light rather than electrons as the data carrier. Light is electrically neutral, and therefore there is no problem with noise and crosstalk. Short wavelengths permit one to regard light as inherently two-dimensional because they enable parallel operations over "pictures," i.e., many picture elements.

At present quantum electronics has very effective lasers for which modulation and control techniques are well developed. Optical fiber communication lines, videodisk memories, and integrated optical logic elements are rapidly appearing. Optical integrated circuits containing injection lasers and connected through waveguides to modulators and other radiation controllers exemplify a radically new path of study whose scientific foundations are, undoubtedly, only coming to light.

Studies of optoelectronic data processing techniques are, indeed, international. Appreciable contributions have been made by researchers in Japan, France, Great Britain, and the Federal Republic of Germany. In the United States this area is actively explored by Bell Telephone, IBM, Stanford University, the University of Pittsburgh, the University of San Diego, and other institutions.

The P. N. Lebedev Physical Institute in Moscow pays much attention to studies of optoelectronics oriented to significant improvement of computer performance. These studies are of an integral nature; designs include multichannel optical communication lines, optical mass memories with fast data writing and erasure, fast digital controlled space light modulators, semiconductor lasers with electronic excitation, and fast scanning

of luminous spots, enabling, for example, the design of address
tubes for data fetch and large-screen high-fidelity color television.

The design of high-throughput optoelectronic processors meets
with numerous difficulties, and although no optoelectronic device
free of essential disadvantages has been proposed to date, many
important partial solutions to this problem do exist.

As hardware for future high-throughput processors is sufficiently developed and is rapidly improving, the time seems to be
ripe for discussions of the structure of would-be optoelectronic
processors, tentative estimation of their performance, and more
precise formulation of requirements for their elements.

In 1974 at the P. N. Lebedev Physical Institute methods were
developed for design of optoelectronic processors operating in a
nonpositional residue number system. At the same time it became
evident that along with high throughput the optoelectronic processor should feature great flexibility, i.e., should execute various tasks with the same high effectiveness. Since 1976 the institute has been working on the design philosophy for optoelectronic
processors based on the control operator method.

The gist of the method is that there should be no dedicated
unit in the optoelectronic processor for logic and arithmetic operations. Data are to be processed in the optical path built of
digital optically controlled space light modulators. Algorithms
are stored in the optical memory in the form of two-dimensional
pictures, and the type of operation is defined only by the type
of operator. This is a promising idea, opening the way to implementation of picture logic.

Such an approach may be helpful in the solution of today's
problems. The method of control operators may be used, for
instance, in the design of channel switches for optical communication lines and various switches for computers, controllers,
self-learning automata, etc.

Analysis of the trends in electronics and optoelectronics shows
that by the end of the 1980s or the beginning of the 1990s electronics will approach the speed of 1 gigaoperation per second or more,
and further throughput improvement will be possible only through
new ideas of which optoelectronics is an example. Fortunately,
many technological processes used by the electronics industry
may be applied either completely or with some changes to fabrication of optoelectronic processors. It seems quite probable that
optoelectronic processors with a speed of 10^{10} operations per second will be designed before the turn of the century.

Academician N. Basov

Contents

I

OPTOELECTRONIC SWITCHING SYSTEMS IN TELECOMMUNICATIONS

1

Fundamentals of Telecommunication Switching Systems

1.1 INTRODUCTION

We are in a period of scientific revolution in considering wired
telecommunications. The technology of transmitting information
by means of manipulating electrons has, from engineering and
economic viewpoints, a considerably different optimization than
do photons of light. The tendency of the 1980s has been to imi-
tate the past. But mimicking electronics can be costly of effort,
reliability, and energy when working with the optoelectronic and
the photonic technologies. Nevertheless, so much effort has been
expended throughout the world that a transition period is being
directed by the administrations of many of the telecommunications
organizations of the world. This portion of the book emphasizes
the transition between science and engineering. The latter now
encompasses optoelectronic and photonic switching in advanced
telecommunications. Advanced electronic and optical telecommunications will undoubtedly coexist, but in differing and changing
degrees. As the enormous advantages become apparent, the
change becomes possible. Thus, certain video switches for the
subscriber loop of telecommunications are inherently expensive,
using electronic switching or repeaters with conversions from
one form of energy to another continuously throughout transmission. Solutions using optoelectronic or photonic technology
often provide more effective results.

To deal with the new functions that must be replaced, or at
least blended, transitions in time and technology must be understood. There may, of course, be cases that do not require any
mixed form of compromise.

The functional requirements of a switching system may, at
first, appear relatively simple and be easily understood, but the

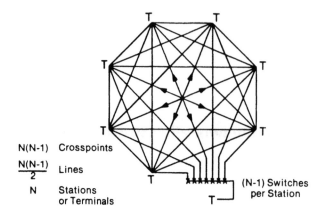

Figure 1.1. Noncentral terminal switching.

service and feature requirements are changing rapidly. With this growth is emerging a new era of telecommunication circuit switching.

There are now two principal segments to such switching, namely, "circuit switching" and "store and forward switching." The former applies to switching where a continuous two-way path is established in space, time, or frequency of the bandwidth that is appropriate to the calling and called station transducers for the duration of the communication. The latter includes the currently popular "packet switching," and usually applies to data transmission where the message is stored in a memory and then transmitted later at a suitable time. In this book, we deal with the philosophical systems design approach and the historical approach. Some specific examples are cited under the classification of the switching field. You are further referred to an excellent review of telecommunications circuit switching by Amos E. Noel, Jr., appearing in the *Proceedings of the IEEE* (Vol. 65, No. 9, September 1977, pp. 1237-1253), entitled "What Is Telecommunications Circuit Switching?"

1.2 CENTRALIZED SWITCHING

Figure 1.1 shows noncentral circuit switching serving n stations or terminals with n(n - 1)/2 two-way transmission facilities or lines. The switches are known as "crosspoints" and complete

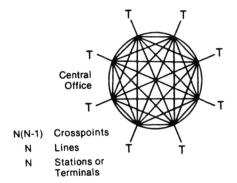

Figure 1.2. Centralized terminal switching, remote-controlled
network.

the transmission paths at both ends of the facilities. Figure 1.1
shows that this requires a total of n(n - 1) crosspoints. When n
is quite small, noncentral systems still have some application. Obviously, centralized terminal switching, as illustrated in Fig. 1.2,
is more efficient, since only a single two-way line is required per
terminal or station, although the number of crosspoints is unchanged. However, the crosspoints can be manually or remotely
operated. Making centralized switching more efficient by halving
the number of crosspoints, we arrive at n(n - 1)/2, or a single-
stage nonblocking switching network (Fig. 1.3). Centralized
switching in nonblocking network form also provides for all stations or terminals to be simultaneously involved in a connection.
When there are less than n/2 simultaneous connections, fewer
crosspoints are required. Figure 1.4 shows that lines, known as
links, can be used within the central office or switching system
for providing fewer than n/2 connections. The concept introduces
a blocking network when all links are in use connecting 2L stations
or terminals, and the remaining (n - 2L) stations or terminals cannot be connected until one of the links is available. The "grade of
service" is dependent upon the number of links L. These concepts
of switching center network topology and teletraffic engineering
are pervasive throughout all telecommunications engineering

1.3 NETWORKING AND BASIC TELECOMMUNICATIONS FUNCTIONS

In public telecommunications, many central switching systems are
deployed, and with a single central office, n lines can replace

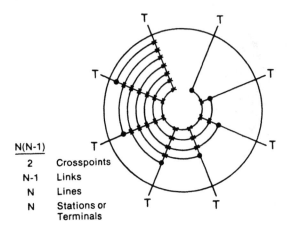

$\frac{N(N-1)}{2}$ Crosspoints
N-1   Links
N   Lines
N   Stations or
      Terminals

Figure 1.3. Centralized terminal switching, nonblocking network.

n(n - 1)/2 lines. This leads, on the average, to fewer and shorter
lines. Trunking consists of connecting one office or entity with
the transmission facilities of another. Thus, more than one trunk
is required, if blocking is to be reduced, to handle the telecommunications load between offices. Collectively, interconnected
central offices, or *nodes*, form the network. Such a transmission and central switching network differs from the in-office link
and switching network (switching center network).
    In practice, a plurality of central offices is usually connected
to an intermediate office rather than directly interconnecting all
the central offices. These intermediate offices are given a series
of names, depending upon the nature of their use, such as "tandem," "transit," "toll," and "gateway." In this usage, central
offices connected to stations are known as *end*, or *local*, offices.
Even intermediate offices may be connected by trunk groups, with
preference given to providing more than one route for a call. Direct trunks connecting two offices are usually selected first, if
idle. Alternate routes, using idle trunks and local and intermediate offices, are more usual.
    There are different methods of routing and connecting, and
this results in different network grids, such as "hierarchical,"
"symmetrical," and "polygrid." Figure 1.5 shows the *ladder*, or
hierarchical public telephone network grid of North America. The
switching plan permits alternate routing up and down the ladder.

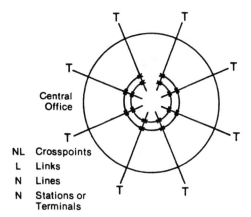

NL Crosspoints
L   Links
N   Lines
N   Stations or
    Terminals

Figure 1.4. Centralized terminal switching, blocking network.

Two types of trunk groups are shown. The *final* groups are between the intermediate offices, or next *class* in the ladder, and
between ladders at class 1 (regional) and higher usage groups,
which permit alternate routine between offices within or between
ladders. In such a network, a particular switching system may

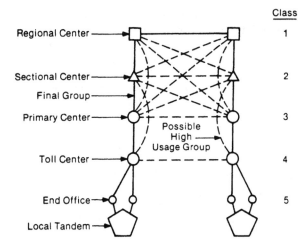

Figure 1.5. North American hierarchical public telephone switching network.

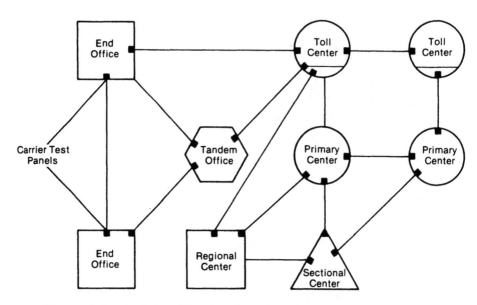

Figure 1.6. An interoffice relationship that also illustrates an advantageous maintenance panel location for intercarrier testing.

perform several functional classes simultaneously and bear multiple names, such as local/tandem, or local/toll.

Figure 1.6 illustrates a maintenance arrangement for prime equipment bays where test and communications requirements are necessary on carrier (or metallic) channels between offices. Such a panel centralizes test equipment in one location for test and maintenance personnel. This figure also illustrates a hypothetical interoffice relationship and its representations.

The basic functions, such as are required by telephony, apply to both manual and automatic switching. Except for special services, manual switching has become obsolete in the United States. The basic functions are divided into three categories: signaling, interconnecting with the link or switching center network, and controlling. They are in the order in which they generally occur during the progress of a call in a late twentieth century switching system in the United States:

Attending: reception by a central office of a request for service from a station or another office.
Signal reception: central office addressing responds to the desired called station.

Interpreting: determining the action required based on the received information.

Path selecting: determining an ideal link or channels or series of links through the switching center network.

Route selecting: determining the trunk group for the path to be established, including interoffice calling.

Busy testing: determining whether a link or trunk is in use or reserved, and, if so, "hunting" for a path by successive testing.

Path establishment: controlling the switching center network to establish an interconnection channel. This circuit switching function requires some form of memory to retain or remember the connections for the duration of the transmission.

Signal transmission: transmitting, in interoffice calls, the address of the call for the distant office.

Alerting: informing the called distant station that it is about to receive a call.

Supervising: detecting when the connection is no longer needed or available for other purposes, as discussed in the following section.

1.4 CONNECTION OR CALL SERVICE FEATURES

Call service features include a numbering plan, call progress tones and announcements, and supervision. In a numbering plan it may be open, closed, or quasi-closed (in North America). Call progress features include dial, busy, ringing, recorder, receiver offhook, and recorded announcements. Supervision features include call abandonment, no address transmitted, partial address transmitted, unassigned address received, and unequipped address received.

In many networks the numbering plan is *open*, meaning that addresses of different length are transmitted to reach a given station, depending upon the calling station location. In the North American public network, the numbering plan is *quasi-closed*. This means that within a numbering plan area there is the same number of digits.

1.5 OPERATION SERVICE FEATURES

Operation service features include provisioning, growth, reliability, management (manual and automatic), operator assistance, maintainability, and charging (e.g., coin and credit).

The degree of blocking in networks is a judgment factor that determines the overall grade (or quality) of service. An administration sets and maintains a grade of service. This usually decides the quantity of equipment required.

Features must be provided in the system to measure the load, type, and activity. Facilities can be designed to obtain traffic measurement data and periodically pass it along to other systems for processing. Switching system reliability and its operation must be taken into account in provisioning. However, to ensure continuity, allowance needs to be made for out-of-service times. Features are designed into systems to enable the system to be maintained and its quality measured. Features are required in the switching system design to recognize requests for, or the need to, connect to an operator, and to provide operators with special signaling and switching features. Thus, there are many considerations in the design and deployment of switching systems that are required, besides those for call processing. Operational service features may, in many respects, resemble modern information processing systems inputs and outputs. They may be magnetic recording units, video, or data links to other information processing systems.

1.6 TELECOMMUNICATIONS SERVICES AND SWITCHING

Telecommunications now include a growing number of public demands for many types of services. Users and residence users may require special services, e.g., business use of another class of switching system known as a private brand exchange (PBX). In the United States, Japan, and France, a service known as CENTREX is offered from the central office. For residence and other users, generally, there is also coin service and shared or party line service, as in the United States and a few other countries. These are but a few of the many services, such as wide area telecommunication service or abbreviated dialing. The definition of a service as distinguished from a feature is that the customer is aware of the feature when using the switching facilities. The number of service combinations that may apply to a given line in some systems may be in the few hundreds. Counting specific features and switching systems designs are required to literally support many hundreds of combinations.

1.7 SWITCHING EVOLUTION

There are many principles in switching. Some are lost due to enormous detail required for their implementation. However, a careful study of switching development reveals certain accepted principles.

Theory provides formalization of the principles. To date, very little theory exists on the structure of switching systems. For the most part, theory has been applied to link and trunk networks and, to a lesser degree, the system control capacity.

To date, the three principal eras of switching, based mostly upon the technology, are manual, electromechanical, and electronic. Initial dates of invention, trial, or commercial service are readily established. A peak year, in terms of number of lines in service, is perhaps one measure, since it establishes the beginning of the decline of an era. Using this definition, manual switching reached its peak about 1938. At present, about 2 to 2.5% of the world's telephone lines are still switched manually. About 6 to 7% of the lines are switched by electronic systems since their start in 1965. Electromechanical switching may not reach its peak worldwide until about 1985.

Although most administrations are currently placing electronic offices in service, it will be a few years before the number of lines served by electromechanical offices reach their peak.

Each of the switching eras, to date, including the electronic era, have made many unique contributions. In this chapter, only the evolution of the basic functions through electronics are covered.

1.7.1 Manual Era

The first switchboards designed for telephone switching were placed in service by 1878. The basic functions remain the same. Much effort was expended during this era to improve the reliability and efficiency of the basic operator apparatus. Of principal importance for densely populated areas was separating the originating and terminating flow of traffic with separate switchboards. Another key concept introduced was the multiple appearance of lines and trunks.

Also developed at this time were the attending and alerting functions, along with the simultaneous development of the telephone line and station. An important innovation and improvement was the centralization of the *common battery*, or battery that could be used to power all station transmitters without cross talk.

1.7.2 Electromechanical Era

Direct Control

The principal objective of the first device and system was remote control of a selective device. The first Strowger two-motion

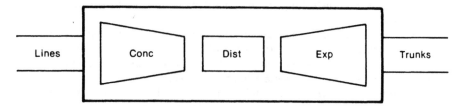

Figure 1.7. Switching network functions.

step-by-step switch was demonstrated in the laboratory in 1879 and produced commercially in 1892. In both cases, they were controlled "directly" from the station by "calling devices."

Gross Motion Switching

The motion required by these mechanisms was supplied by electromagnetically operated ratchets. Because of space traversed between terminals, switches of this type are classified as "gross motion." A two-motion switch size then emerged that was 10 X 10, which fitted the decimal numbers then being assigned to telephone addressing. Within 5 years, a switching center network design evolved that was interconnected by links, providing access to more stations. A series of three stages of switches could for the first time be used to reach one of 10,000 terminals. This was known as "progressive" step-by-step switching.

The other stage, called the *line finder*, made it possible to reduce the total number of switches. Thus, each line did not need a full series of 1 + 10 + 100 = 111 switches to reach 10,000 terminations. The selection stages were arranged serially to reach any line. The links between stages were shared. The switching center network therefore takes on the more general internal functions shown schematically in Fig. 1.7.

Indirect Control

Interconnecting was the one function that had been the focal point of technological change until the electromechanical era. By 1906, it was proposed that nondecimal selection stages would be possible if the customer dialed into a storage device. Decimal selectors with capacities as much as five times (500 terminals) the step-by-step selector were devised. Sending pulses were required only during the addressing phase. A concentration switching network

using the same or other types of selectors was inserted between
the line concentration and the first selector stage. Upon selection completion (or caller abandonment), the sender could then
be released for reuse on other calls. Thus an element of control
was divorced from the connection path, and the genesis of indirect control began.

As the number and speed of pulses to set a selector increased,
more accurate pulsing (address signaling) methods were developed.
Fully automatic systems of this type were placed into production
about 1920. Later the indirect control principle was extended,
and it became possible to further translate the digits dialed to
designate the central office. The central office code digits referred to such plurality as *translators* or *decoders*. The holding time of a translator could be only a few hundred milliseconds
per call. Thus arose the first application of bulk memory in information processing.

Fine Motion Switching

The logic required for sequencing and storing information in control systems was carried out by electrical relays. Such relays
were used even for crosspoints. Switches with coordinate contact arrays served as a replacement for the panel switch, with
the advantage of "fine" rather than gross motion, and, as compared with a matrix of relays, did not require a magnet or coil
for each crosspoint.

The crossbar switch was the first coordinate switch to use this
principle in combination with reliable contacts.

Common Control

During the coordinate switch development, the idea of using simple crosspoints without individual switch controls began to evolve.
The concept was to provide a common circuit that would control
the operation of selected crosspoints and then be available to select and operate other crosspoints for other calls. This was the
origin of the network control access network (NCAN). A supervisory circuit can hold the activated crosspoints electrically for
the duration of the call. (This is a "memory" of the connection.)

From this grew the "coordinate link" principle illustrated in
Fig. 1.8. In such an arrangement, each input can reach many
outputs in the second stage. These are also accessible, with
blocking, to the other inputs. Such a principle may be extended
in many ways (even optically, as will be seen in Chap. 3). It
was then possible to free the entire network control process from

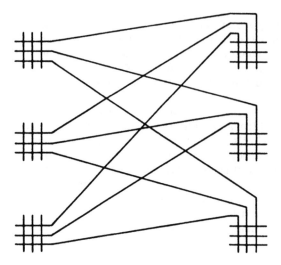

Figure 1.8. Coordinate link principle for larger selection.

the progressive establishment of connections through the switching center network. As a result, an idle trunk to a suitable destination, including possible alternate routes, can be chosen before selecting the switching center network path. On terminating
calls, "look-ahead" gave other advantages. It made possible busy
line determination of the called line, or to selection of an idle line
in a PBX group, without establishing a connection from the incoming trunk. Also, it enabled the placement of switches on the
lines, according to traffic requirements, and independent of their
directory numbers. (A translator or bulk memory indicates the
correspondence between directory and "equipment" numbers.)
A switching center network consists only of crossbar switches
and associated NCAN multicontact relays.

1.7.3 Signaling Development

Signaling is generally divided into associated and disassociated
functions. Initially, direct current was used for loop, composite, and simplex signaling. Other associated signaling included
supervisory and pulsing (as in dialing). Alternating current
signaling includes supervisory, pulsing, and digital, as in per
channel and common channel. Disassociated is normally common
channel use only.

The signaling function takes place from stations to the switching system, in which case it is known as *station signaling*. Between switching offices, it is known as *interoffice signaling*. More recently it has become possible, with the introduction of digital transmission facilities, to allot separate bits for signaling.

From an historical point of view, a new generation of disassociated signaling, known as *common channel signaling*, is beginning to reappear. Common channel signaling uses modern data transmission channels to interchange supervisory and address signaling generally between processor-controlled switching systems.

1.7.4 Electronic Era Developments

Reed relays were a product of World War II development and use hermetically sealed contacts in a glass envelope, which are inserted into electromagnetic coils. They were used to avoid troubles that could occur in some open-contact crosspoint devices, such as crossbar switches.

Time-Division Switching

Another development from wartime transmission and theory is time sharing, or time multiplexing. Sampled speech signals could be multiplexed over broadband channels. Inside the switching systems, broadband links were easier to apply than over distances. Furthermore, the sample, if digitized, may be transmitted over parallel channels at lower pulse rates. The concept of switching speech samples with electronic crosspoints that are simultaneously shared by other calls is called *time division*. In more recent times, digitized or coded samples have been switched.

The invention of the transistor, a solid-state device with no appreciable standby power, has been found to be an excellent technology for direct use in switching.

Electronic Switching Center Networks

Electronic crosspoints are used in electronic switching center networks. In space-division switching, memory is required. In electronic space-division networks, the memory is usually built into the crosspoint electrically so that it will be sustained once it is activated.

In time-division switching center networks, a cyclic memory is separate from the crosspoint. During the cyclic readout of each address of the memory known as a *time slot*, the calling and called terminals are connected together long enough for the transference of the required information. Network capacity may be extended at least two ways. One is the storage of the samples in either analog or digital form, independent of a time slot. The other technique is a space-division network that is switched at a rate necessary to provide sufficient time slots. It has become known as a *time-multiplexed switching* (TMS) *network*. The first method, or time-slot interchange (TSI), permits the sample to be placed in the same, or another medium, in a different time slot from the memory.

Time-Division Control

Time-shared electronic logic with a common electronic memory made it possible to reduce the complexity of control accessibility. In particular, input signals may be examined one at a time in a process known as *scanning*, and the control functions as a whole, including scanning, are scheduled serially, or time shared.

Stored Program Control

In 1955, a new flexible type of stored program control for switching systems was introduced, using general-purpose electronic logic and bulk memory. Memory was used for the control logic, as well as for the translations to temporarily record other information about a call.

By removing the system logic from the hardware of the control, a new degree of design independence became possible. The common control, or *central control*, became the heart of general-purpose processors for switching systems. The controls are general-purpose and time shared, and the input and output functions related to call processing are also time shared and general-purpose. The output information is conveyed by "distributors," and the input information accessed by scanners. This is collectively called distributing and scanning function (DAS).

Design emphasis of system control has switched from hardware to software, particularly by continued design after the system was installed. With more recent developments in large-scale integrated circuits, read-only memory (ROM) with programmed logic are now used for the logic needs that remain in many systems. This class of systems is called *distributed logic systems* and still requires a central control for directing or sequencing the information flow. These peripheral functions of centrally controlled systems can also be described as almost autonomous units making possible another approach to distributed control. The processors may be used as a plurality of common controls, as in electromechanical systems. This division of functions is related to program selections rather than just to the peripherals.

The deployment choice among full stored program control (FSPC), distributed control, action translator control, and wired logic is dependent upon the objective of the installing and operating administration. At present, distributed control system designs are in a state of flux, so that although the intentions of some designers are to retain stored program control flexibility in the central control, others are distributing the functions and, with it, the flexibility.

1.7.5 Optoelectronic Era Development

This is the subject of Chaps. 2 through 4.

1.8 SWITCHING DEFINITIONS

Here, *circuit switching* can be defined as providing through connections for the exchange of messages. Circuit switching can also be defined as applied to a system with a switching center network used for two-way transmission with no chance of transmission delay due to passage through the network.

Obviously, the switching center network may be implemented to carry any type of information delivered to it in electrical form. It is also possible for the control portion of the system to operate simultaneously with more than one switching center network. The transmission may represent different types or speeds of messages, and as such, several types may be carried through the same switching center network simultaneously.

The other generic form of switching applies, in general, to one-way transmission that may be delayed. This is known as *store and forward switching*. This generally implies that no switching center network is employed, but as indicated above, a signaling network is required, and the storage aspect of this type of switching can vary widely with different types of applications. The message itself may also be divided into smaller, usually uniform, sized segments. This form of store and forward switching is known as *packet switching*. In this case, each packet must be preceded by a repetition of the called address.

1.9 GROWTH OF SWITCHING SERVICES AND FEATURES

After the introduction of each new generation of switching technology, considerable effort is expended in adapting it to new features and offering new services. This may extend over many years and reach its peak before proliferating another technology. The feature and service development process varies widely around the world, as do the general areas of applications of the switching system, PBX, local central office, intermediate office, store and forward, etc.

The high-speed interchange of information between offices made possible by the application of common channel signaling (and now optoelectronic signaling and transmission) will bring to the entire public network of telecommunication offices the same advantages stored program control brought to the switching offices themselves.

2

Optoelectronic Switching in Telecommunications

2.1 INTRODUCTION

In Chap. 1 the overall present status and historical progression
of the fundamentals of telecommunications switching were out-
lined. In this chapter we focus on the fundamentals of electronic switching and the development of optoelectronic switching at the device and microcircuit levels. Concentration is placed
on that part of electronics and optoelectronics that carries over
into optoelectronic switching designs of the present, and perhaps
the next two decades — in other words, useful engineering technical and historical guidelines for innovating further development
at the device and integrated optical circuit levels.

2.2 EVOLUTION INTO PRESENT AND FUTURE MICROCIRCUITRY

Very large scale integration (VLSI) has evolved to a point where
well over 10,000 digital gates are fabricated on a single integrated
circuit (IC). The first generation of microelectronics evolved in
the 1950s. At that time, discrete transistors, diodes, and passive components, including thick- and thin-film circuits on printed
circuit boards, led to lower power, lighter electronic systems. In
the 1960s, the integrated circuit led to a second generation of
microelectronics on a single wafer of semiconductor material. This
contained the functional equivalent of discrete transistors, diodes,
resistors, and capacitors, and their interconnections. In such
small-scale integration (SSI), a single monolithic IC could be designed to function as a complete digital logic or a monostable mul-
tivibrator. With greater elemental density in the 1970s came a third
generation of medium-scale integration (MSI) and then large-scale
integration (LSI). This evolution to very large scale integration

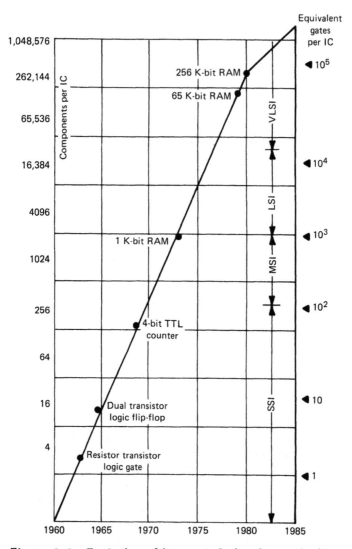

Figure 2.1. Evolution of integrated circuit complexity.

is shown in Fig. 2.1. The complexity is the number of transistors, the number of memory bits, or the number of digital logic gates contained on a single chip. Currently such chips contain the equivalent of 500,000 transistors. The 1980s has seen the introduction of VLSI and very high speed integrated circuits (VHSIC).

The metal oxide semiconductor field effect transistor (MOSFET) has a very high input impedance and pentodelike characteristics. Small in size, with low-power dissipation, inherent high reliability, and mechanical ruggedness, the MOSFET offers virtually complete isolation between the input and the output.

The most effective application of the MOSFET appears to be in VLSI technology. Lower cost, fewer processing steps, and greater functional density are the advantages of MOSFET IC over bipolar IC. The first MOS IC was developed in 1963. MOS large-scale integrated circuits, employing thousands of transistors, have been commercially available for some time.

CMOS is being developed for a variety of low-power and very high speed IC applications. Silicon-on-sapphire (SOS) CMOS inverter structure is expanded, in many designs, to form thousands of digital gates on a single IC.

Microelectronic chips that contain nearly a half-million circuit elements and are hundreds of times faster than currently available devices are being developed. The first VHSIC chips were fabricated using photolithography and have device geometries as small as 1.25 μm. In the mid-1980s, chips are made with electron-beam lithography and have device geometries of submicrometer dimensions. Included in this class of specialized applications that require high speed are levels of radiation hardness that cannot be attained with silicon or SOS microcircuits. Gallium arsenide IC are being developed for specialized LSI-VLSI applications.

Gallium arsenide (GaAs), silicon, and germanium are materials useful for fabricating transistors and integrated circuits, as is silicon on sapphire. There is now a trend toward gallium arsenide for selected high-speed requirements.

The speed of a circuit is measured in terms of the number of operations per second performed by the circuit. To increase the speed, the transistor size can be reduced and higher speed materials used. Gallium arsenide IC, with propagation delays ranging down to less than 50 psec, are being developed. This improves the speed performance of both silicon and SOS IC. GaAs IC are fabricated with a different device structure and manufacturing technology than those used for silicon IC. If a high-resistivity semi-insulating dielectric substrate is used (e.g., 10^7 to 10^9 ohm-cm) as host substrate for interconnection metallization, the stray capacitance is much lower, meaning smaller resistor-capacitor (RC) time constants and an improvement in speed performance.

Since optical devices, such as laser diodes, can be fabricated from GaAs and can be integrated on the same chip with logic functions, the use of optical fiber transmission techniques and other applications currently are being developed. GaAs field effect transistors (FET) offer superior performance as low-noise devices at L-band, S-band, and C-band microwave frequencies.

Device design and device processing represent technical areas where different competing approaches are establishing a substantial technology base, particularly in the interactions of level integration and lithography.

As integrated optical techniques develop, they will also substitute for bulk devices, as in electronics. By way of example we show the appreciable power savings by going to integrated optics as opposed to bulk optical devices.

To recognize the power savings of an integrated optical waveguide modulator relative to that for a bulk polarization modulator, we compare the power required as given for a bulk modulator by Kaminow and Turner in 1966. This gives the power dissipated in driving a modulator as

$$P = \frac{V^2}{4} (C + C_a) \, \Delta w = \frac{1}{4} \left(\frac{\lambda \eta d}{\pi n^3 rL} \right)^2 (C + C_a) \, \Delta w$$

where C = capacity associated with the modulator
$\quad$ = $\varepsilon_0 \varepsilon A / d = \varepsilon_0 \varepsilon \, dl / d = \varepsilon_0 \varepsilon L$
$\quad d^2$ = area of input and output face of square modulator
$\quad L$ $\quad$ = length of modulator
$\quad \eta$ $\quad$ = modulation index
$\quad \Delta w$ = modulator bandwidth
$\quad C_a$ = stray capacity

For an integrated optical modulator (IOM) where 10 V may be used for 100% modulation, and calling $C \gg C_a$,

$$P_{IOM} = \frac{10^2}{4} \times \varepsilon_0 \varepsilon \times 1 \text{ cm} \times \Delta w = \frac{100}{4} \, \varepsilon_0 \varepsilon \, \Delta w$$

In the case of the bulk polarization modulator, the light beam to be modulated passes through the electrooptic crystal in a straight line. Considering a Gaussian laser beam as propagating in the lowest-order transverse mode and focused to just pass through a cylinder of length L and refractive index n, we can calculate the power required for the bulk modulator.

The diameter of the cylinder is a minimum when L is equal to the confocal parameter of the beam. In such a case the beam diameter is $2w_0$ at the waist of the beam and $8^{1/2} w_0$ at the ends of the cylinder, with $w_0^2 = \lambda L/2\pi n$. For cylinder of diameter d equal to $S \times 8^{1/2} w_0$, where S is a safety factor greater than or equal to 1, the minimum value of d^2/L is $S^2 \times 4\lambda/n\pi$. Assuming $C \gg C_a$ and a cylinder of diameter approximately d by a square of dimension d on each side,

$$P_{bulk} = \frac{1}{4}\left(\frac{\lambda\eta}{\pi n^3 r}\right)^2 \frac{d^2}{L} \frac{1}{L} \varepsilon\varepsilon_0 L \; \Delta w$$

$$= \frac{1}{4} \frac{\lambda^2\eta^2}{\pi^2 n^6 r^2} \frac{S^2 4\lambda}{n\pi} \frac{1}{L} \varepsilon\varepsilon_0 L \; \Delta w$$

for a wavelength $\lambda = 1.15 \times 10^{-4}$ cm and $\eta = \pi$ (100% modulation) $n^3 = 6 \times 10^{-9}$ cm V, assuming also S = 6 for easy alignment and L = 1 cm (the same as the integrated optical modulator), we obtain

$$P_{bulk} = \frac{1}{4} 750^2 \; \varepsilon\varepsilon_0 \times 1 \text{ cm} \times \Delta w$$

or

$$\frac{P_{IOM}}{P_{bulk}} = \frac{10^2}{750^2} = 1.8 \times 10^{-4}$$

Thus, the integrated optical modulator-switch requires about 1/5000 the power of the bulk modulator. This illustrates the enormous efficiency of integrated optical modulators-switches over bulk devices.

2.3 SIMPLE OPTOELECTRONIC SWITCHES

2.3.1 Introduction

In this section we discuss optoelectronically controlling the flow of light in optical waveguides. We confine our attention to switches, although some switches and modulators are capable of performing both operations. Optical communications requires efficient devices of either or the two combined. As indicated in the previous section, using bulk methods, the cost and power requirements may be relatively high as compared with conventional communication systems operating in the VHF (very high frequency) range or higher. High drive power can arise due to the relative weakness of available effects for switching and modulating light. Bulk devices require relatively large-volume devices with strong electric, magnetic, or acoustic fields. The stored energy is high, and the

drive power is proportional to the stored energy. Optical waveguides can make use of weak physical effects and often have at
least one dimension of the order of an optical wavelength. Savings
in drive power can be one or more orders of magnitude.

An optoelectronic modulator is a device that alters a detectable
property of a light wave, such as intensity, phase, polarization,
and wavelength, in response to an applied electrical signal. Switching usually refers to the operation by which the spatial location of
a light wave is changed in response to an applied electrical signal
(or an applied optical signal in the case of certain photonic devices). Switching may imply spatial or angular separation between the switched and unswitched beam. Amplitude modulators
can perform this operation and include acousto- and electrooptic
grating modulators, various mode-to-mode couplers, and wave-
guide-to-waveguide couplers.

One important figure of merit of a switch is the degree of cross
talk or the isolation achieved between two or more locations.
Switching time is often shown as related to bandwidth by

$$T = \frac{2\pi}{\Delta f}$$

The specific energy, in most applications, is often less important than the power required to keep the switch in an on
mode. The insertion loss of a switch is also an important figure of merit.

2.3.2 Fiber Optics Switching Methods

The switching of multimode fiber light is a challenging problem
whose solution is being pursued along several avenues. Currently, among the chief approaches under development are the
electromechanical, magnetooptic, and electrooptic. The electrooptic category includes both liquid crystal devices and solid-state
$LiTaO_3$ devices. The electromechanical switch is probably closest
to commercial realization, but the liquid crystal switch, which is
a recent development, is probably a more versatile device.

Switching Functions

An optical switch sends optical signals from one optical fiber to
another in response to electrical commands. Two of the most basic
and useful switch configurations are the 2 X 2 and 1 X N multipole
switches.

In a fiber optic network, switching can be performed optically
or electrically. The optical approach has the advantage of elimi-
nating optical-to-electrical and electrical-to-optical conversion

devices, thereby enhancing the system's security and reliability, and eventually reducing overall system costs.

Two of the main applications for switches are multiterminal data bases and packet switching networks. Packet networks have a complex system of fiber transmission lines that come together at various nodes. By placing switches at nodes, one can select various optical paths through the network on a dynamic basis, with rapid updating for message switching.

Electromechanical Switches

Bell Labs has made a mechanical four-way optical switch. It employs a hermetically sealed flexible tube with a square internal cross section. Switching is with a solenoid, which bends the tube and forces the input fiber to one of the four corners to couple with one of the four output fibers.

With the tube filled with index matching fluid, insertion loss averaged 0.25 dB; cross talk was less than 60 dB. The tube was mounted in a robust case. The switch performed 1 million cycles without failure. Single-mode fiber versions have been made that have losses between 0.8 and 1.5 dB.

A 4 X 4 optical switch made up of hexagonal glass blocks that slide in and out of parallel light beams was made by Nippon Telegraphic and Telephone Public Corporation (NTT). Cross talk was less than -55 dB, and insertion loss was 0.3 to 1.3 dB, depending on the number of block elements needed in the signal's path to obtain the desired switching. The switching elements were moved by electromagnets, with 18-msec response time. It should be possible, using the same principles, to make 100 X 100 optical switches with less than 10 dB total insertion loss, the maximum allowable loss according to considerations of switching theory. Such devices, however, are all very expensive.

Solid-State Electrooptic Switches

Optical switching devices in single-crystal solids are usually based on Pockel's effect, a linear shift in refractive index with applied voltage. The voltage-induced index change is limited in practice to one part in a thousand by the maximum E field that can be applied without dielectric breakdown. Switching speeds are inherently very fast, of the order of nanoseconds (RC limited). At one point in time, research focused on $LiTaO_3$, which offers the best combination of optical quality, high resistivity, large-scale availability, and freedom from optical damage.

Generally, in the $LiTaO_3$ devices, it takes 400 to 500 V to switch a substantial portion of the input fiber light to some other fibers. Several electrooptic switch possibilities are available as products and fiber optic network use and design. The basic materials are either lithium tantalate or liquid crystal, in a number of configurations, to achieve various subscriber or switching center functions.

The lithium tantalate is basically more expensive and uses higher voltages. The liquid crystal types are millisecond switching and can be used for packets or messages. The lithium tantalate is microsecond or even several nanoseconds in switching. Insertion losses can be of the order of 1 dB operating at 30 to 50 V with -18 dB or less cross talk. Designs can include an optional bypass for repeaters at low loss, and 1 X 4 or 1 X N switches, or even a 3 X 4 fiber optic switchboard. Coupling is usually by microlenses of the Selfoc cylindrical type or other types of refractive gradient lenses. As the sensors are multimode, use is readily achieved physically. Active taps or couplers reduce losses in a CATV system as compared with fixed taps.

Magnetooptic Switching

Magnetooptic switching is a somewhat more recent promising technique for fiber switching that has not been too extensively exploited because of the quality of magnetic materials available. This situation has changed with the advent of improved materials.

Switching devices use the efficient optical diffraction from magnetic stripe domains in thin films of iron garnet materials. The width and angular orientation of the uniform parallel domains are influenced by an externally applied magnetic field; thus, the optical diffraction angle and azimuth angle can be controlled. Diffraction is independent of the input polarization.

Such switches have both memory and microsecond response. Domains point in the direction of the last field application, so the switch remains indefinitely in any azimuthal state without power. Microsecond switching speeds are possible due to rapid domain formation and high domain mobility. In practice, the speeds are limited by control circuit inductance.

Since each technique has a fairly distinct set of characteristics, the choice of fiber switch is highly dependent upon the fiber system requirements. For example, if switching rates of 100 MHz are essential, fast electrooptic devices, such as $LiTaO_3$, may be a choice, although high voltages are then needed. (Mode-to-mode and fiber-to-fiber coupling may also be used.) A need for memory could lead to magnetooptics, and a requirement for low-voltage control with microampere drive currents would mandate liquid crystals.

Table 2.1 Examples of Multimode Fiber Optic Switching Techniques

	Electro-mechanical	Electrooptic LiTaO	Electro-optic liquid crystal	Magnetooptic
Optical inser-tion loss	∿1 dB	5.7-12 dB	∿1 dB	4 dB minimum in theory
Switching speed	msec	nsec	msec	μsec
Optical cross talk	-40 dB	-26 to -12 dB	-48 dB	-40 dB, est.
Reliability	Has moving parts	High	High	High
Control power requirements	2-4 V, 0.1-0.2 A	400-500 V, A to mA	5-30 V, 20 A	2-4 V, 0.1-1.0 A

The electromechanical and liquid crystal devices are competitive in several respects. At this time, the electromechanical
switch is a marketable item because of packaging developments.
Liquid crystal devices can be packaged in the same manner, with
four or more fiber pigtails emanating from a switch box, as can
switched couplers. Table 2.1 lists some various properties of
fiber optics switching techniques.

2.4 SWITCHING WITH PHOTODIODES

2.4.1 Introduction

Optoelectronic switches have been shown to have isolation in excess of 80 dB and can readily operate between a few and several
gigahertz. This section discusses its principals, models of switching in PIN and avalanche photodiodes, and potential applications.
Such a method avoids problems of directly routing the optical
carrier. Devices for switching bound waves in optical waveguides
may also be used, particularly with electromechanical optical fiber
switches. These often suffer from problems of speed and reliability, and nonmechanical optical waveguide switches often have low
isolation and too high an optical insertion loss. Optoelectronic

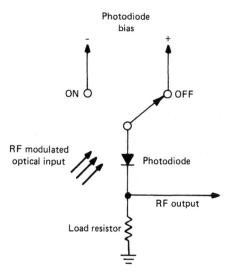

Figure 2.2. Operating principle of an optoelectronic switch (off state of the switch).

switches, which are a hybrid of electronic and optical switching, can handle high-frequency signals at high performance into the GH$_z$ range.

2.4.2 Optoelectronic Switches

The optoelectronic switch is usually in array form, in which its input and output ports appear purely electronic. The principle is demonstrated in Fig. 2.2. The signal, which may be transmitted as an intensity-modulated light beam, when intercepted by the photodiode, acts as a photodetector and crosspoint switch element. By reverse biasing the photodiode, the optical signal is converted into an electrical signal that may then be detected across the load resistor shown. Detection across the load resistor represents the on state of the switch. By forward biasing, the sensitivity to the optical signal is greatly reduced (off state). The light optical signal power may come from a passive element, such as optical waveguides, optical waveguide power dividers, or a freely falling light beam, as from a lens or lens array. Switching is achieved by converting the optical signal back into an electrical signal. Switch elements can be configured,

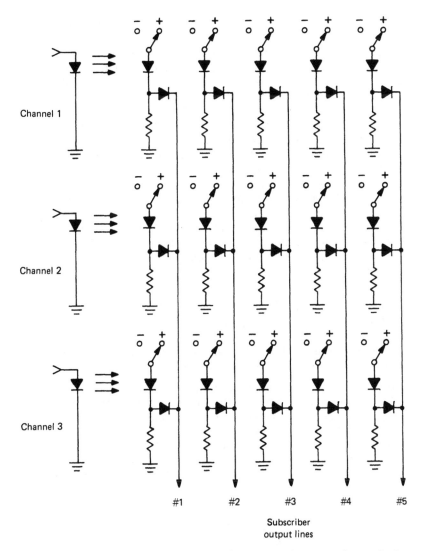

Figure 2.3. Conceptual schematic for a 3 X 5 crosspoint switch matrix.

as shown in Fig. 2.3, to make a multielement crosspoint matrix switch. In this figure, a matrix switch selectively routes three channels of information to five subscribers, for example. The switch shown consists of three 1 X 5 crosspoint modules, one for each channel of information. Laser diodes or light-emitting diode sources may be used. In using optoelectronic switching, photodiodes have a reduced response in the forward bias rather than the reverse bias condition usual for photodetection. The isolation I can be defined as the power ratio of the electrical signal at the load during reverse bias, as opposed to forward bias, for a given optical signal incident. Isolation is also dependent upon the signal frequency, power, and wavelength of the optical carrier. Forward bias can reduce the power of the electrical analog of the optical signal detected at the load by three independent processes: (1) reduction of the internal impedance of the device from a high value at reverse bias to a low value at forward bias; (2) in photodiodes with avalanche gain, there is the loss of gain under forward bias (switching of avalanche gain is independent of the signal frequency); and (3) the direct reduction of the quantum efficiency of the photodiodes in forward bias (due to heavy injection in the junction). MacDonald and Hara (Ref. 2) have shown the isolation I to be

$$I = \left(\frac{\eta_R}{\eta_F}\right)^2 G^2 \frac{|Z_{DR}|^2 |Z_i + Z_{SF}|^2}{|Z_{DF}|^2 |Z_i + Z_{SR}|^2}$$

or the ratio of the received electrical power P_e when the diode is reverse biased when it is forward biased

where η = quantum efficiency of the photodiode

  Z_s = source impedance of the photodiode and its load

  Z_D = internal impedance of the photodiode under the bias conditions employed

  R,F = subscripts appropriate to reverse and forward bias

  Z_L = load impedance used in conjunction with the photodiode

  G = charge multiplication factor of an avalanche photodiode is used

  Z_i = input impedance of the external observing instrument

$$P_e = \left|\frac{Z_L V_o}{Z_L + Z_D}\right|^2 \frac{Z_i}{|Z_i + Z_s|^2}$$

$V_o = \frac{hc}{e\lambda} \eta P_o Z_D$ = voltage across the load

P_o = incident signal power

Optoelectronic switching arrays appear to have no exact counterpart in purely electronic apparatus. They reduce leakage of signal and have no reflections back into the power divider as a result of various switch states. Large numbers of optoelectronic

crosspoints can be supplied from a single source of signal power, as no attenuation must be included in the power division
network to suppress reflections. Optoelectronic switches require two power needs. One is sufficient power to maintain
the desired signal-to-noise ratio, and the second is sufficient
electrical power to maintain the required forward current in
the off condition. Thus, average signal powers may be of the
order of 10 µW to maintain a video channel signal-to-noise ratio
of 40 to 50 dB with a typical PIN photodiode, with even less
power for avalanche photodiodes. Optoelectronic switches using
avalanche photodiodes can readily switch at frequencies over
10 GH_z.

2.5 PROMISING BISTABLE OPTICAL SUBPICOSECOND SWITCH-
ING DEVICES

2.5.1 Introduction

Many all-optical devices have been demonstrated: memory, limiter,
ac amplifier, oscillator, gate, modulator, and discriminator. In
current optical systems, transmission may be optical; regenera-
tion, amplification, and modulation are still largely electrical.
Fewer light electricity transformations should be necessary as
all optical devices evolve. It becomes natural to turn to semiconductors in search for optical bistability.

For optical data processing, practical devices can evolve to
be submicron in size with speeds in the picosecond range requiring little switching energy, such as 10^{-15} W and holding
powers of less than the milliwatt range. Operation can be at
room temperature, and they can be incorporated into integrated
systems (see Ref. 3).

Bistable optical devices in one form are passive optical devices
that would exhibit two or more states of transmission of light
switchable between these states by a temporary change in the
level of light input. The output versus input characteristics
exhibit optical hysteresis. Memory elements, differential amplifiers, pulse shapers and limiters, optical triodes, and logic elements are but a few of the applications for bistable optical devices.

2.5.2 Features of Bistable Optical Switching Devices

The four principle features include (1) their ability to process
light directly; (2) a capability for parallel processing; (3) a

very large bandwidth, e.g., greater than 10^{13} Hz; and (4) an ultrashort switching time. Such devices can be superior in bandwidth and switching time to electronic or even Josephson switching devices. They meet the need for high-capacity repeater and terminal systems for processing optical signals. In addition, in two-dimensional array form, they have the ability to operate directly on a signal already in light form.

An important parameter of an optical material is its optical length (physical length multiplied by refractive index). In turn, the optical length is proportional to the optical phase shift that a light wave undergoes when passing through the material. To obtain bistability, two simultaneous conditions are required:

1. The optical length within the nonlinear material depends on the intensity of the light in the resonator.
2. Conversely, the intensity of the light in a resonator depends on the optical path length.

Because of the nonlinear material, the optical path length in the resonator is proportional to the light intensity within it.

As bistable optical switching devices can have an optical hysteresis curve, they can be useful for an optical memory. Within the hysteresis region there are two possible values of stable output power for a constant input power. Assuming constant input power, an additional "switch-up" power pulse places the system in a high power output state. Subsequently, a reduction of power below the "switch-down" switches the system to the low power condition.

For amplification of a small modulation, the relatively steep output versus input characteristic is used. It amplifies a small modulator of an input carrier beam. The gain can be varied by changing the resonator tuning of the system. The same characteristics can be used for pulse shaping. Thus, an input power less than that required to reach the steep slope region is transmitted weakly. Weak signals can thus be discriminated against. A large change in light transmission through a bistable optical device can be produced by a small change in resonator tuning. If this tuning is provided by a secondary optical beam, a weak optical signal can be used to control the transmission of an intense light beam. This optical "triode" or "transistor" uses a relatively weak signal beam to impose information on (modulate) a strong continuous wave (carrier) beam. The power-limiting characteristics of the optical bistable switching device may also be used to stabilize the peak power of pulses in an optical communications system. Further, with large optical nonlinearities, many interlocking hysteresis

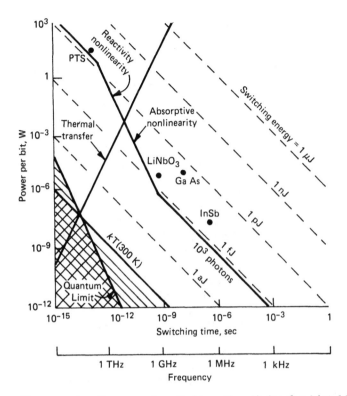

Figure 2.4. Power and switching time limits for bistable optical devices. (Based on Ref. 4.) *Note:* PTS is polymerized diacetylene.

loops may be obtained. This type of characteristic makes these devices useful for analog to digital conversion or multilevel logic operations.

2.5.3 Switching Speed and Power

The wide range of switching powers and speeds of bistable optical devices is shown in Fig. 2.4. The fastest devices require the highest switching powers. The lowest switching energy shown is for a hybrid integrated optical bistable device. The segments delineate limits. The frequency scale applies for repetitive switching. Quantum mechanical considerations indicate a switching operation must dissipate at least h/t energy, where h is Planck's constant and t is the switching time. The thermal transfer range is based upon heat dissipation during each switching operation and restricts the achievable switching rate. Thermodynamically, a yes-no switching operation must dissipate a minimum of kT of energy, where T is the absolute temperature and k is Boltzmann's constant.

REFERENCES

1. J. M. Hammer, in *Topics in Applied Physics, Volume 7, Integrated Optics* (T. Tamir, ed.), Springer-Verlag, Berlin, 1979, 139-200.
2. R. I. MacDonald and E. H. Hara, *Switching and Photodiodes*, IEEE J. Quantum Electron., March 1980, 289-295.
3. H. M. Gibbs, S. L. McCall, and T. N. C. Venkatesan, *Optical Bistable Devices: The Basic Components of All-Optical Systems?*, Optical Engineering, July/August 1980, 463-468.
4. P. W. Smith and W. J. Tomlinson, *Bistable Optical Devices Promise Subpicosecond Switching*, IEEE Spectrum, June 1981, 26-33.
5. H. M. Gibbs, S. L. McCall, and T. N. C. Venkatesan, *Optics News*, Summer 1979, 6-12.
6. D. Smith and D. Miller, *Computing at the Speed of Light*, New Scientist, February 21, 1980, 554-556.
7. R. W. Keyes, *Physical Limits in Digital Electronics*, Proc. IEEE, May 1975, 740-767.
8. C. M. Bowden, M. Ciftan, and H. R. Robl (eds.), *Optical Bistability*, Proceedings of the First International Conference on Optical Bistability, Plenum Press, New York, 1981.
9. IEEE Journal of Quantum Electronic on Optical Bistability, Vol. QE-17, March 1981.

3

Optoelectronic Network Design and Hardware

3.1 INTRODUCTION

To introduce the subject of network design, this book first considers the implementation of data communication networks and later encompasses voice, video, and data as a whole. By initially taking a user's perspective we attempt to present a rather full menu of possible approaches and the arguments for and against each approach (see Ref. 1).

In data communications at the close of the twentieth century, the prime concerns are with binary facilities and transmission between nodes or locations. Between two permanent nodes this may be point-to-point or switched. The use of the term *facilities* refers to the communications channels and equipment taken in its entirety.

A network consists of three or more nodes in various configurations, such as star, loop, and hierarchial, as discussed in Sec. 3.3. Thus, a *star* network is a centralized network, and a *loop* network encompasses the distributed concept. A further consideration is the way the network is utilized. In a calling mode, or circuit switching mode, there is a high degree of interactive communications between modes, such as two individuals talking to each other over a public telephone network. Message switching presents messages to the network that usually have a high degree of channel utilization but a low degree of interaction. The introduction of packet switching in the 1960s led to nodal computers to process the packets of different messages. This has led to packet-switched services. Systems that are built by adding on to other systems tend to break down after several iterations of add-ons, largely because the original objectives of the systems were altered. Designers and design strategies should consider potential objectives as well as actual ones.

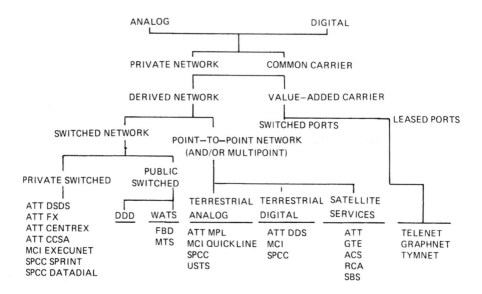

Figure 3.1. A decision tree for considering alternatives to networking in the United States.

Establishing a value to the benefits derived by a data communications system may involve such internal considerations as

1. Expected decrease in overhead costs relative to revenue generation
2. Overall increase in gross revenues
3. Increase in gross revenue per system employees

What are the time values of present and projected items of information? This key question relates the manner in which delay of information affects an organization's revenues. A decision tree is often convenient to ensure that all important considerations have been examined. Such a tree is illustrated in Fig. 3.1.

In certain countries, such as the United States, the first question to be resolved is whether to construct a wholly owned private communications system rather than utilizing common carrier service. This requires a number of considerations, such as

1. Geographical distribution of locations
2. Availability of rights of way
3. Availability of communications employees

4. Availability of communications equipment and construction
   permits
5. Availability of operating licenses from the proper agencies

   The benefits of a private system versus that of a common carrier system have to be balanced. A common carrier system typically has at least the following advantages.

1. No large capital investment is required.
2. Widespread geographic coverage is available immediately.
3. Redundancy is generally available for backup in case of breakdown.
4. Communication services are available immediately.

   A private system provides cost stability in system operation
and user control of system services, and the economics of scale
can readily be achieved. An assessment of the state of technology and rate of obsolescence is a major factor, as equipment and
plant obsolescence can be rapid in the last two decades of the
twentieth century. Private communications are often attractive
in single-area locations, such as an office building, plant or fac-
tory, or university campus. Not too many organizations can staff
a private communications system, but there is now a growing tendency for large United States companies to do this.

   Another key question is the choice of analog or digital facilities. Obviously, for data communications, digital data communications systems are generally superior for signal conversion and error control. Where choices exist in a hybrid system, the economics usually favors the analog system, as a system error rate is
only as good as the worst link error rate, and the digital link
then does not contribute as much to the system. The trend is
to move toward digital, but the currently installed base is largely
analog.

   Another key design strategy question is whether to use derived networks or value-added carriers. This often concerns the
nature of the traffic between node locations, the system control
philosophy, and system redundancy requirements. A derived
network is one in which the different nodes are specified by the
system designer. A value-added common carrier refers to carriers with specialized networks that accept digital messages and
pass these messages to their destination node via intermediate
nodes. This leads to considerations of traffic patterns and
whether the traffic is independent, to a great extent, from central control. In a central control, a star network can be set up
as a derived network and the message switching device is at the

central node. With increased message volume the system response time can increase exponentially to an unacceptably high value. In such a case it may pay to establish direct communication links between high-traffic volume nodes. This leads to decentralization of system control. It is, however, possible to reduce the number of required communication links by having some nodes function as message switches. Also, if the network is a hierarchical system with centralized control, the derived network could well be the choice. This implies that, from an economics point of view, the costs may be larger for the value-added carrier because of the overhead of network control.

When time value is high, system redundancy may be required. This leads to additional expenses for a derived network, since common carriers may have considerable redundancy. Backup communications facilities are a design must.

For small numbers of messages transmitted occasionally, switched facilities are normally more economical. As traffic increases, point-to-point communications offer an economic advantage.

The decision tree shown in Fig. 3.1 illustrates the alternatives of various services. A tariff study of the rate structure of the different services is very helpful, particularly if costed out to a similar base for charges per unit time (per month or hour) and the amount of unused time. A possible evolution of common carrier-based networks is shown in Fig. 3.2.

A formal decision process can be developed for an organization planning to implement a telecommunications network. One of the most readily implemented formal procedures is for a data network.

3.2 EVOLUTION OF PRESENT OPTICAL NETWORK SWITCHING

3.2.1 Introduction

As fiber optic and other optical light-wave technology has advanced, so also has data communications in packet switching, voice digitization techniques, and the development of the integrated voice, video, and data switching network. This required a reevaluation of issues and trade-offs and analysis of problem areas in the design of optical communications switching networks. This section examines the impact of fiber and integrated optics on switching network architecture in terms of their current and future cost effectiveness. In particular, the present integration of voice and data into the same common user network has given rise to an integrated switching network concept that is promoted

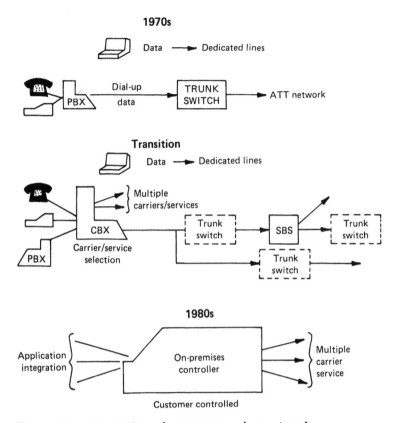

Figure 3.2. Evolution of common carrier networks.

by the established common carriers in the United States. These carriers have matched the switching concept to traffic characteristics.

Circuit switching has been shown to be more cost effective for traffic characterized by long contiguous messages, and packet switching more cost effective for traffic with short messages. Basically, also, it appears that switching methods should judiciously match the proper switching strategy to a given traffic type. The optimum switching strategy is very dependent upon the end-user requirements.

3.2.2 Point-to-Point Optical Links

Optical frequencies may employ modulation and coding techniques used in electrical baseband and carrier transmission systems, as well as wave division multiplexing. One of the most common systems for present optical links is binary on-off modulation with square-law detection.

The light sources used are typically injection lasers or high-radiance light-emitting diodes (LED). The dominant noise limitation using a simple photodetector followed by a baseband amplifier is the thermal noise of the amplifier. The use of an avalanche photodetector (APD) reduces this thermal noise effect, and the dominant noise is then the excess shot noise.

In optical transmitters it is usually desirable to match the radiation emission area to the waveguide dimensions. Modulating current drives is a common practice, with drive currents typically of the order of 100 mA, achieving an output of several milliwatts for both sources. As the LED usually radiates in a directional pattern that is broader than the laser, only about 0.05 mW is coupled into an optical fiber. A laser diode, on the other hand, usually radiates over a narrow angle that may be better matched to a fiber's acceptance angle, and powers may be coupled that are in excess of 1 mW. As the laser threshold varies with temperature, feedback control is often used to prebias the laser just below the threshold to obtain a fixed peak output in the on mode. The maximum distance between repeaters may be calculated when the loss per unit length of lightguide, the power coupled into the lightguide, and the receiver threshold for a given bit rate are known.

3.2.3 Network Design Considerations

With the wide band capability of optical fiber transmission systems, data bus distribution networks are readily achievable where many spatially distributed terminals are served with the same multiplexed signal. Thus, subscriber terminals in telecommunications can take on computer architecture concepts. However, a data bus using optical fibers is quite different from metallic wires and cables in many aspects. For example, the techniques and components for coupling signals are quite different, and the performance and construction of a data bus depends, to a major extent, upon the way terminals manage signal transactions on the bus.

Two configurations, discussed further in Sec. 3.3, are the serial distribution system using access or T couplers, and a parallel distribution system using a star coupler. Such distribution systems can supply nonblocking access among all subscribers without the use of conventional switching nodes.

In data bus design two major considerations are (1) the maximum data channel loss between any pair of stations, and (2) the maximum dynamic range that needs to be allowed for communication between a station pair. In such cases, the star coupler has an advantage over an inline configuration. Using a star coupler with uniform power distribution to all input and output ports, the dynamic range is due to the attenuation differences of the fibers connecting the coupler to the terminals. On the other hand, a star configuration may require a greater total length of fiber cable than the inline design.

With a large number of subscribers, some form of light-wave switching will probably be effective to conserve cabling and interconnect devices. Such switching also provides for more flexible routing, traffic, and service features.

In areas where access switching was formerly used to conserve transmission, bandwidth may no longer be necessary due to the greater reduction in the cost of bandwidth provided by light-wave technology. The less expensive bandwidth also enhances the use of wide-band digitized voice techniques. This also enhances security and such additional service features as facsimile, video services, and computer communications.

As fiber optics goes into the existing common carrier transmission plant, it will operate with the switching nodes already in place by using some type of conversion at the transmission/switching interface or by switching node modification to accommodate direct optical or optoelectronic switching.

3.2.4 Some Optical Switching Developments

One of the most cost effective ways of implementing direct optical switching of information on optical waveguides is through the use of integrated optics (Ref. 2). Such devices used in conjunction with single-mode optical fibers provide high data rate switching matrices that can be designed to operate up through the tens of gigahertz range. Electronic switches are also limited by being governed by the transmitted data rate; changes in state for the optical switch are controlled by the switching role, which is typically two or three orders of magnitude lower than the data rate. The power dissipation of integrated optics is potentially lower than that of electronic switching. Signal reduction from an integrated optical switch reduces severe cross-talk problems experienced in microwave at high data rates.

Table 3.1 Some Classes of Integrated Optical Switches

Intraguide mode conversion types, using

1. Forward, or codirectional
2. Backward, or contradirectional
3. Angular (refraction, reflection, Raman-Nath scattering, Bragg diffraction)

Interguide mode conversion types, using

1. Forward, or codirectional
2. Backward, or contradirectional
3. Angular

Relative phase modulation of orthogonal modes, mode interference

Mode attenuation, or loss modulation (absorption or scattering types)

Guided-to-unguided mode conversion, unguided-to-guided mode conversion

---

Optical waveguides and hollow metallic waveguides have some common features. Both can have a finite number of guided modes at any given frequency, and both can have mode conversion problems due to imperfect geometry. Metallic waveguides can usually only support guided modes. Thus, mode conversion is limited to the interchange of power among a finite number of guided modes. Optical dielectric waveguides, on the other hand, also possess a continuum of unguided radiation modes. The optical switching process transforms an incident optical mode into a new mode. This leads to substrate modes, radiation modes, and guided modes with differing polarizations. Optical waveguides and the surroundings are thus characterized by dielectric and magnetic permeability constants that permit a great deal of flexibility in devising a variety of designs, for example, for unique switch designs. Using external control fields of a electrical, acoustic, or magnetic variety may perturb the dielectric and permeability constants to obtain switching action. Several classes of possible integrated optical switches are shown in Table 3.1.

In principle, one way of configuring optical switching matrices is to make an array of optical waveguide switches, e.g., a 2 X 2 matrix, as illustrated in Fig. 3.3. These may be interconnected by dielectric waveguides on a single substrate and the number

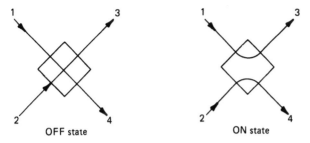

Figure 3.3. An example of an optical crosspoint type of switch in a 2 X 2 array.

minimized, as pointed out in Ref. 3. Such analyses usually assume crossings among transmission paths that are allowed by interconnect switches. However, noncrossing networks can also be formed. For the noncrossing network an efficient algorithm requires that the state of each switch be chosen so that its output signals are in the same relative position in the waveguide matrix that they are required to occupy in the output state. Where N is greater than 3 for an N X N network of this type, N(N - 1)/2 switches and N switching stages are required. Figure 3.4 shows an example of an 8 X 6 connecting network that can be constructed from 2 X 2 optical switches interconnected by dielectric waveguides on a unit substrate.

Coupling such switches to optical data transfer networks presents a problem of the polarization state of propagating waves because of connecting fibers to relatively flat microoptical devices (Ref. 4). As both polarizations are often present in

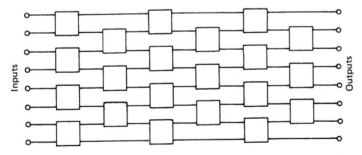

Figure 3.4. An example of an optical switch using an array of 2 X 2 switches to make a 6 X 8 architecture.

single-mode fibers and since integrated optical thin film devices often possess different polarizations with different efficiencies, special design techniques must be used to avoid the degradation of performance.

3.3 PROMISING OPTICAL NETWORKING AND SWITCHING DEVELOPMENTS

3.3.1 Introduction

Optical communication devices have been shown to facilitate the distribution of tasks and programs among a considerable number of relatively slow machines (Ref. 5). Such "second generation" approaches often trade off band width and low self-induced noise for simpler and more flexible program distribution and task implementation. These so-called universal systems are basically designed for fiber optics rather than substituting fiber optics for a previous metallic cable system. Some of the more cost effective advanced systems utilize metallic cable and optical waveguide technology. Ring and common bus topologies appear the most suitable for implementation utilizing optical channels. Typical current loop systems range from a 20 to a 200 Mb/sec capability (Ref. 6). The growth of microprocessor applications satisfies this need to communicate with all or some dissimilar machines in a given network. This also permits flexibility in adding and deleting machines. When one compares common buses and loops, the latter are often preferable from power budgeting, interface requirement, network modifiability, and protocol design points of view. Chapter 4 describes other universal optical network and local area networks.

3.3.2 Network Structuring

A local network may consist of nodes interconnected by terminals, switches, and links. A switch is not necessarily a hardware device; it can be a conceptual term that denotes a switching operation. Thus, copying the content of a register into another register performs a switching operation.

Optical channels may be used to interconnect nodes that optimize communications in the following ways.

1. The numbers as switches are reduced, and the switches themselves are simple.
2. The interface complexity between the network and the terminal is minimized.

3. The computational effort that each terminal requires is minimized.
4. Dissimilar terminals may be contained in the network.
5. Terminals may be added or deleted without requiring network modification.
6. The links basically have no band-width limitations.

Several typical topologies are the star, common bus, loop, and fully connected networks. Their general characteristics in relation to the criteria cited above are given as follows (see Refs. 7-10).

I. Star network (Fig. 3.5):
   A. Link numbers: N.
   B. Termination numbers: 2N.
   C. Switch numbers: 1.
   D. Switch complexity: N.
   E. Modifiability: Switch must be modified.
   F. Remarks: The link bandwidth must be equal to or greater than the maximum transmission rate of the fastest terminal. A single point of failure is possible in a complex high-speed switch.

II. Common bus network (Fig. 3.6):
   A. Link numbers: N.
   B. Termination numbers: 2N.
   C. Switch numbers: N.
   D. Switch complexity: 2.
   E. Modifiability: Simple.
   F. Remarks: The link bandwidth must be greater than or equal to the maximum transmission rate of the system's fastest terminal. A single point of failure is possible.

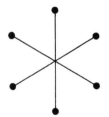

Figure 3.5. Star network.

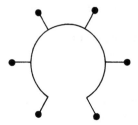

Figure 3.6. Common bus network.

III. Loop Network (Fig. 3.7):
    A. Link numbers: N.
    B. Termination numbers: 2N.
    C. Number of switches: N.
    D. Switch complexity: 2.
    E. Modifiability: Simple.
    F. Remarks: Some delays are required at each node. The
        link bandwidth must be greater than or equal to the
        maximum transmission rate of the system's fastest ter-
        minal. A single point of failure is possible.

IV. Fully connected network (Fig. 3.8):
    A. Link numbers: $N(N - 1)$.
    B. Termination numbers: $N/2(N - 1)$.
    C. Switch numbers: N.
    D. Switch complexity: $N - 1$ (switch positions).
    E. Modifiability: Difficult.
    F. Remarks: The reliability is high. The link bandwidth
        is equal or greater than the maximum transmission rate
        of two connected terminals.

Figure 3.7. Loop network.

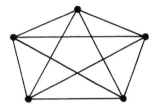

Figure 3.8. Fully connected network.

3.3.3 Common Buses and Loops

For the types of applications discussed above, the loop system appears to have some advantages over the common bus topology.

Protocols

In both loops and common buses, contention, token passing, or time-division multiple access can be provided. In the loop system, advantage can be taken of a node delay and means of accomplishing this in a simple manner have been described in Refs. 11 and 12. The latter is not necessarily so in common bus systems.

Cable Requirements

Present common bus structures at times require more cable than a loop for the same location (Ref. 11).

Power Requirements

The minimum received power at the receiver corresponds to a transmitted power $\bar{P}_t \geq NP_r + P_{NL}$,

where P_r = power at the receiver = $P_t/N - P_{NL}$
$\quad\quad\quad P_t$ = transmitted power
$\quad\quad\quad P_{NL}$ = insertion loss of the cabling and couplers, which is not necessarily equal for all the receivers
$\quad\quad\quad N$ = number of terminals

assuming a given bit error rate at a given bandwidth. If the network requires reconfiguration, the lower bound on the transmitted power is

$$\bar{P}_t \geq N_{max}P_r + P_{NL}$$

Thus, in a topographic system using a loop structure, the transmitted power requirements are

$$\overline{P}_{tloop} > P_r + P_{Nloop}$$

where P_{Nloop} = link losses due to the cable and connectors. (*Note:* $\overline{P}_{tloop}$ is independent of N.)

In the case cited, the bus topology has higher power requirements and possibly lower reliability, as lasers may be required, as well as an upper bound on the system's number of allowed terminals.

3.4 PROMISING OPTOELECTRONIC AND PHOTONIC COMPLEXITY ANALYSIS DESIGN APPROACHES

3.4.1 Introduction

This section illustrates a number of design approaches that have been used, and some illustrative design approaches, as well as simple hardware illustrations. The relative hardware costs, of course, are shown in the applicable time frame and geographic location. They do, however, indicate the types of considerations that may sometimes be required.

3.4.2 Previous Optical Network Topology

The first wired city fiber optics network, Hi-OVIS 1, used a design that was conservative and appropriate for the first experimental installation. This conservative design, shown at the top of Fig. 3.9, uses one optical fiber for downstream signals and one fiber for upstream signals. Each subscriber terminal was equipped with a keyboard, television, camera, and microphone to permit videophone communication and interactive services with the head end. The 168 subscribers were provided with 30 channels, using a fixed-matrix video switch configuration. Transmission was all analog, using LED, and receivers used PIN photodiodes. This design, which was appropriate for a conservative installation, required two fibers per subscriber from the head end to the home terminal, two pairs of transmitters and receivers per subscriber, and a 2 X 30 video switch for each terminal at the head end.

The expansion of the Hi-OVIS concept, shown at the bottom of Fig. 3.10, reduces the fiber requirements by using wave-division multiplexing (WDM) between the subscriber terminal and the

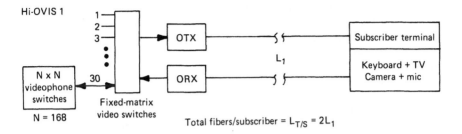

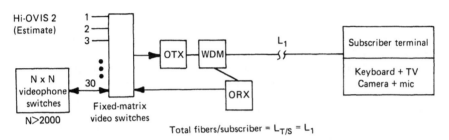

Figure 3.9. Hi-OVIS network topology.

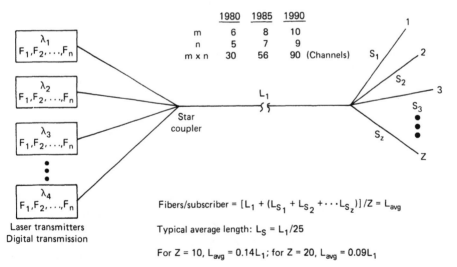

Figure 3.10. Star coupler optical network topology, center to subscribers.

head end. The rest of the system is essentially the same, with the exception of using various options for terminal equipment depending upon location. For example, business terminals may use several "slave" television sets connected to the terminal receiver, or various types of facsimile equipment.

The inherent problem with most optical network topologies currently used and proposed for wired cities is that they are expensive. The expense comes from overdesign and lack of utilization of the unique properties that fiber optics has to offer, compared with wire and coaxial designs.

The large systems proposed or used in Japan, Canada, and France are government funded and established in part to develop components and technology to improve the social and economic welfare of the nation. These systems are rather expensive, as currently designed, to permit cost-effective large-scale duplication and implementation on a national level.

To achieve economic feasibility means to offset installation, operation, and maintenance costs with subscriber revenues. To attain high revenue levels, an interactive network must provide a large number of services, achieve a high penetration of subscribers, and minimize average installation costs. Typical wired city designs provide CATV channels, videophone, high-fidelity channels, and perhaps videotext. To increase the subscriber revenues for the head-end system operator, other services should be offered. Strong possibilities here include

Tiered pay television systems, in which subscribers pay an additional charge for special services, or the ability to watch special TV movies or programs.

Fire, police, burglar, and medical alarm services and energy management services, connected to the head-end computer and local stations.

If nations aspire to export interactive fiber optic broadband networks, designs must permit a reasonable cost per subscriber.

Anticipated Cost Reductions

To reduce the costs of fiber optics networks, several major cost items must be reduced. These major items include optical fibers, optical transmitters, and switching requirements. The cost of optical fibers is, of course, expected to continue to drop over the next decade. The costs of light sources and detectors are expected to show the most dramatic decreases over the coming decade. Switching in these interactive networks still represents a problem, which should continue into the late 1980s, until low-cost optical switches can be developed.

It is obvious that a reduction in fiber prices, particularly light source prices, will help to reduce the overall systems costs for fiber optics wired cities. However, examination of the projected installed costs per subscriber, even using lower cost fibers and light sources, still does not make such systems economically attractive for large-scale commercialization. Further dramatic cost reductions must be developed and used to make such large networks economically viable.

The best way to achieve rapid cost reductions is to design the overall system to require less equipment, fewer components, and lower installation (labor) costs. The Hi-OVIS 2 design shown makes an attempt at this by cutting fiber requirements per subscriber in half, but even here the average installed cost is still too high for large-scale usage.

Further technological achievements permit different effective optical network topologies. For example, such achievements include

1. Low insertion loss WDM splitters
2. Low insertion loss transmission star couplers
3. Electrooptic tunable filters (EOTF)
4. Low-cost video (digital to analog, D/A) and (analog to digital, A/D) converters

In addition to current technological achievements, the star coupler design must take into consideration that wired communities are intended for gradual expansion and implementation, which will occur over a span of several years. The entire fiber optics and electrooptics industry is in a state of rapid development. New components are being developed and readied for sales and mass production, and new research efforts are exploring new techniques, particularly for optical switching and optical repeaters. The star coupler design plans for the incorporation of new components and devices over the next few years. In particular, long-wavelength laser diodes and APD are planned for implementation. The overall design is also flexible enough to handle new technological advances in optical switching, should they become available toward the latter part of the 1980s.

The major accomplishments of the design are

1. An 85% reduction in the amount of optical fiber required, as compared to the Hi-OVIS 1 design
2. A 30-fold (or more) reduction in the number of switches required at the centers or subcenters

3. Potential expansion of channels from the original 30 to 90 on installation by the year 1990
4. A 30% (or more) reduction in the number of transmitters required per subscriber

Center to Subscriber Transmission

The illustrative network topology for transmission from centers or subcenters to subscribers is shown in Fig. 3.10. The basis for this portion of the system is the combined use of star couplers, WDM, and FDM (frequency-division multiplexing) using laser transmitters. The individual laser transmitters are mostly used for dedicated downstream transmissions of noninteractive CATV channels and FM radio (hi-fi) channels.

The number of transmitters required for each subscriber is a function of: (1) the number of wavelengths utilized in the WDM transmission system, (2) the number of frequency bands used in the FDM schemes, (3) the length of optical fiber between the transmitters and receivers, (4) the insertion losses of the star couplers, (5) the coupled optical power from the laser source, (6) the sensitivity of the PIN or APD used in the subcenter terminal, (7) the insertion loss in the subscriber terminal (from the connector and EOTF, electrooptic tunable filter), (8) the attenuation of the fiber at the shorter wavelengths, and (9) the pulse dispersion over the spectral range utilized, and other factors.

The small chart in Fig. 3.10 indicates a number of expected wavelengths and frequencies that might be used in 1980, 1985, and 1990. To achieve the required number of channels, more frequency bands or more wavelengths can be used. The practical limit on wavelengths depends upon the use of active or passive filters at the receiver, the required spectral separation between adjacent channels, the use of temperature controllers on lasers and detectors, and other factors. In 1990, the use of 10 wavelengths may be considered conservative, compared to 20 wavelengths that could be technologically accomplished. Further developments in EOTF technology would be required to increase the wavelength number above 10. The thrust of this design is to minimize the required number of transmitters, and thus reduce the overall system costs. The practical limit on the frequency domain is mostly determined by the trade-offs in optical power. The more frequency bands used, the lower the peak optical power of each channel. Given possible expected transmission lengths of about 1 to 5 km and expected fiber attenuations less than 3 dB/km, five to nine frequency bands seem to be a reasonable compromise. More frequency bands means that fewer subscribers can be put onto the star coupler at the receiving side.

Since wire communities are planned for an extended growth period, more wavelengths can be added to handle further channel requirements at some future period. The long-wavelength region in particular may be more economically used in the near future, whereas the shorter wavelengths are available now. Future expansion to additional subscribers in a small local area may be accommodated by leaving one or two available fibers at the receiving star coupler to connect to future terminals. This is similar to telephone companies leaving unused wire pairs in typical cable installations.

Whereas the number of laser transmitters for each downstream fiber may be six to twelve, the number of subscriber terminals served by that fiber may be much higher. This is one of the major cost attractions of this design. As shown in Fig. 3.10, the accumulation of insertion loss using a star coupler is relatively gradual. The star network is ideally suited for a large number of terminals tapping off one fiber. As depicted in Fig. 3.10, a total of 10 terminals represents a loss of 18 dB and 30 terminals less than 24 dB.

Using laser transmitters and digital transmission, a reasonable optical power margin for video transmission is 50 dB, meaning a 50 dB loss from the transmitter to the receiver. Using commercially available optical fibers and laser diodes, the optical power into the star coupler at the receiving side should be -15 dB or higher. With a 50 dB power margin, that leaves 35 dB for the star coupler. The power margin assumes using APD in the receivers and includes all safety factors and conservative derating requirements on components.

The large power margin for the star coupler means that 20-45 subscribers could be connected to the transmission fiber. The more subscribers using the one downstream fiber, the greater the economic savings. Assuming only 20 subscribers are connected to the star coupler, the average length of fiber per subscriber is only $0.09L_1$, where L_1 is the distance from the center to the average subscriber terminal. This calculation is shown at the bottom of Fig. 3.10.

For the conservative case of 20 subscribers using one downstream fiber, a significant reduction in the total required transmitters is accomplished. If each subscriber has one upstream transmitter for videophone and interactive services, it means that $(20 + 6)/20$, or 1.3, transmitters per subscriber are used. This

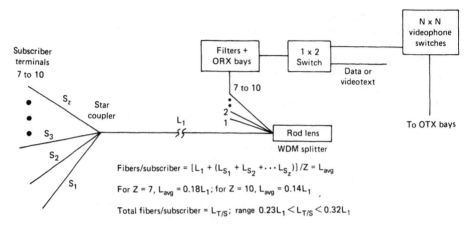

Figure 3.11. Star coupler optical network topology, subscriber to center.

is significantly less than the two transmitters required per subscriber for the Hi-OVIS 1 design.

Subscriber to Center Transmission

For the optical transmission of subscribers to the local subcenters, analog transmission may be used for the first installations using narrow spectral width LED. Figure 3.11 shows a possible network topology for subscriber terminals to the subcenter. Again, use is made of star couplers to connect seven to ten subscribers, each using a different LED wavelength to the upstream fiber. The subcenter has a passive WDM splitter to divide the incoming wavelengths into separate receivers. The possible WDM demultiplexer here is a graded index lens with an integral blazed grating.

The optical receivers used for each subscriber could have detectors that utilize narrow-bandpass thin film coatings on the semiconductor surface, to achieve better channel isolation and lower interoptical channel cross talk. These coatings are specifically made for WDM designs and permit LED to be used with relatively narrow spectral spacings. Seven wavelengths may be used now, and perhaps nine wavelengths by 1985, for this upstream LED transmission. Once low-cost A/D converters are available, upstream transmission would be best done by lasers using more wavelengths and more coupled power.

The optical receivers for each subscriber could then connect to the equivalent of a 1 X 2 switch (transistor) to divide video-phone requests from the other interactive services. The video-phone signals could then be connected to a seldom-blocking switching bank and to a downstream optical transmitter. One or two units of the downstream transmitters should be totally dedicated to video telephone transmission, to simplify the head-end engineering.

Since fewer subscribers are connected to the star coupler for upstream transmission, the savings in optical fiber per subscriber is not as dramatic as in the downstream case. Here, as shown at the bottom of Fig. 3.11, the fiber required per subscriber is about $0.25\ L_1$. The major savings using this design for upstream transmission is reduction in switching requirements. This example shown in Fig. 3.11 uses a 1 X 2 switch, compared with the 30 X 2 fixed-matrix switching bank per subscriber used for Hi-OVIS.

A Subscriber Terminal Example

The components used for the subscriber terminal depend upon many factors. Some of the major criteria used to determine terminal components include (1) offered services and system channel capacity, (2) use of video telephone and/or auxiliary facsimile devices, (3) analog versus digital transmission from the head end, and (4) other factors.

One possible configuration for a subscriber terminal for the illustrative network topology is shown in Fig. 3.12. The transmission from the subcenter is assumed to be digital, and telephone service is used by or available to the home subscriber.

The depicted terminal uses an EOTF controlled by a microprocessor. The coding to the microprocessor comes from the keyboard; numbers on the keyboard correspond to channel numbers. Each channel number has a specified wavelength, which is tuned in by the EOTF. A feedback loop from the broad-band APD receiver can help to keep the wavelength peak centered, by slight voltage changes in the EOTF contacts. The keyboard signals also control frequency filters, to pick one electrical frequency band from one optical wavelength.

The tapped electrical signal can then go to a video D/A converter connected to the television set. The channel being viewed can be shown by a simple two-number LED display. Each component requires electrical power at different voltages, which could be supplied from a centrally controlled ac/dc power supply. The

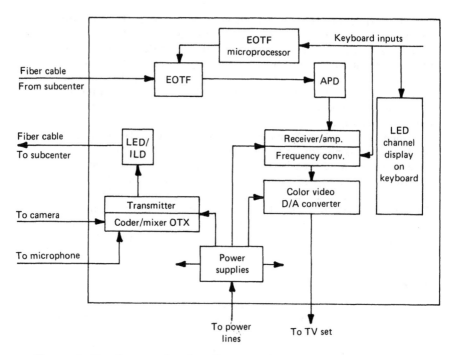

Figure 3.12. Proposed subscriber terminal block diagram.

upstream transmission is one using a LED transmitter connected
to the videophone camera, microphone, and perhaps a telephone
or facsimile device.

Potential Design Improvements

Many refinements and improvements can be added to the basic illustrative design. One area for further development is in the control, coding, connection, and operation of the switching network.
This contains several elements or levels of complexity. The presence of the video telephone makes the switching design more complicated than that currently used for telephone narrow-band
switching systems.

As described earlier, the future of telecommunications will see
more and more covergence between computers and communications
equipment. This, in part, includes a trend toward more software
and less hardware for any given link or network. One way to

reduce switching components requirements and optical transmission equipment is to use dynamic software programming for switching controls. An example of such dynamic programming is as follows.

During start-up and initial operation, the head-end computer can keep track of video telephone and interactive service calls made by subscribers. Some channels end up in use more than others. Some subscribers communicate frequently with some other subscribers, but almost never to others.

After evaluating channel distribution to small local areas or to individual subscribers, the center or subcenters can implement small changes in the switching connections. These changes would more evenly distribute traffic to reduce waiting times and incidences of receiving busy signals for videophone calls. Note that the illustrative design does not provide one laser transmitter wavelength and frequency allocation per subscriber, and that waiting or queueing will occur. This suggested dynamic switching programming will greatly reduce average waiting times and incidences of busy signaling.

Another example of dynamic programming would be to change the channel allocations on the transmitter, depending upon usage. If some interactive functions, such as videotext, are used extensively by some subscribers but not by others, redistribution of channels will again reduce waiting times for the retrieval of required information. This concept of dynamic programming and nonfixed switching matrices is in sharp contrast to previous designs, which require expensive switching equipment for each subscriber.

A further area of improvement could also be for a broad wavelength region APD design. Using only one detector may be difficult, especially for expanding the wavelength region from 800 to 1600 nm. A split-level or hybrid APD detector may eventually be developed to give superior response over the complete wavelength region. Diminished responsivity at longer wavelengths is compensated for by decreased fiber attenuation.

Illustrative Design Summary and Economics

The illustrative design for the wired city optical network topology represents a step forward in the realization of economical optical interactive broad-band systems. The savings accomplished in this design include substantial reductions in the optical fiber required per subscriber, as well as the number of transmitters used in the overall system. Further savings are achieved by greatly reducing the required switching components and connections required at the head end or subcenter.

A summary of the basic equipment requirements for four wired city designs is given in Table 3.2. This table compares the basic savings in fiber, switches, and transmitter and indicates the technology and components required. Hypothetical cost estimates for studying Hi-OVIS 1 are given in Table 3.3.

To obtain definitive estimates of the cost reduction of the illustrative design compared to the Hi-OVIS designs is difficult. This requres price estimates for the WDM splitters, EOTF components, and transmission star couplers. The cost of each of these is, of course, dependent upon the quantity manufactured. Each of these components is unique to the illustrative design, as compared with the other wired city designs. Fortunately, each component has a market outside wired city network designs, and can be used in many other fiber optics applications. It is reasonable to expect that the costs for these items will follow the same sort of price versus quantity curves as do laser diodes. Thus, a star coupler or WDM splitter, which may be expensive in sample quantities, could be made in mass production of units of 100s or 1000s for less than 1/20 the cost. The costs of the EOTF components are more difficult to estimate, since packaging in the manufacturing process is a major cost factor. In quantities in excess of 1000, it is reasonable to expect these components to drop to a level below $100 each in 1983 U.S. dollars.

Other items difficult to consider in definitive cost comparisons between the illustrative design and others are labor costs and time estimates. The illustrative network uses very little fiber compared to other designs, which means fewer fiber cables, fewer splices, and less required installation time. The design also uses fewer transmitters, which reduces labor requirements for head-end installations. The degree to which all these items will affect total labor costs is difficult to estimate.

Other factors that influence cost estimates are the number of subscribers to be connected to the star couplers. The greater the number of terminals operating from a star coupler, the lower the total system cost. In practice, there will be a distribution of star coupler density or fibers per coupler, ranging from perhaps 7 to 40, depending upon local subscriber density. Hotels and apartment buildings will be most cost-effective to connect to this optical network, since local subscriber density is relatively high. Private homes in more sparsely populated areas will be more costly to connect to any optical or coaxial system.

Table 3.2 Basic Equipment Comparison for Wired City Projects[a]

Project name	Fiber/ sub- scriber	Switches in centers	Optical transmitters	Optical receivers	Special devices comments
Hi-OVIS 1	$2L_1$	(30×2)/subscriber + $N \times N$ videophone, $N = 168$	2/Subscriber (LED), (A)	2/Subscriber (PIN)	Expensive fixed-matrix switching/subscriber, 30 channels, no subcenters
Hi-OVIS 2 (possible design)	L_1	(30×2)/subscriber + $N \times N$ videophone, $N = 2000$	2/Subscriber (ILD + LED), (A + D)	2/Subscriber (PIN + APD)	2 WDM couplers/subscriber, advanced center services, 30 channels, 3 subcenters
Illustrative design (star coupler)	$0.3L_1$	(1×2)/subscriber + (SB) videophone, $N = 2000$	1.3/Subscriber (ILD + LED), (A + D)	2/Subscriber (PIN + APD)	1 WDM Demux./10 subscribers, 4 star couplers/10 subscribers, uses video A/D/A, WDM, EOTF, 30 channels expandable to 90, low-cost switching

[a] A = Analog transmission; D = digital transmission; L_1 = average distance from subscriber to subcenter; SB = seldom-blocking switching system.

Table 3.3 Hypothetical Hi-OVIS 1 Cost Estimates[a]

Equipment item description	System requirement	Cost
Optical transmitter and receiver units	2 Pairs/N	4000N
Fiber cables from subcenters	2X/N	2XFN
TV + keyboard + controller +	2-Way/N	4000N
Video switches and power	2-Way/N	2600N
Head-end computer controller	1 Installed	$300,000
Subscriber distribution and drop boxes	Cable dependent	90N
Installation cost of all fiber cabling	100 labor-hour/N	100LN
Head-end and subcenter fiber distribution	Drawer and racks	150N
Fiber connectors installed	2 Demountable/N	60N
Installation cost of head-end connections	Equipment and interface	30LN
Permanent station test equipment (E + O)	Meter, scopes, etc.	$40,000
Portable test and installation equipment	Splicing and measurement	$50,000
Video cassette recorder/players and tapes	10 Installed	$25,000
Automatic cassette changers and cassettes	4 Installed	$60,000
Complete still-picture system units	1 Installed	$25,000
Color character generator system	1 Installed	$15,000
TV retransmission equipment and antennas	Channel dependent	$360,000

Table 3.3 (continued)

Equipment item description	System requirement	Cost
Color monitor TV sets in floor racks	30 Installed	$25,000
Complete head-end and local studios	Central location	$1,150,000
Professional fees, license, shipping, taxes, engineering, travel, etc.	Variable	$1,950,000

[a]X = average subscriber link distance from subcenter = 1800 m. C = number of available service channels = 30. L = labor charges per labor-hour = $10.00. F = cost of fiber in cables per meter = $3.00 (1977 prices). N = number of subscribers. Using above variables and values, the estimated cost equation is: Cost = $4,000,000 + $23,000N ($N$ = number of subscribers).

3.4.3 Advanced Switching Methods

Switching for wired communities or applications in which numerous terminals are interconnected can be a very expensive part of the total installed system. Switching can be accomplished optically or electrically or by any hybrid combination. In most previous systems and many currently under construction, the predominant means of switching is to convert the incoming optical signal to the electronic domain, and then use traditional types of electronic switches. These electronic switches for optical networks usually are computer controlled. Full-matrix video switches for broad-band applications and digital TDM (time-division multiplexing) or SDM (space-division multiplexing) switches have been used in optical networks. However, given conventional fiber optics designs, these switches and associated hardware and software became a major system expense. Costly switching raises the installed price per subscriber to a level that currently cannot be justified by increased revenues from broad-band optical networks.

Several types of switches are currently required for optical information networks for large communities. These switches are mostly determined by the types of services offered by the

network, such as two-way data services, CATV and audio channels, and videophone. The most costly switching requirements
are for the video telephone services. This situation is expected
to change, to permit videoconferencing for business as well as
home subscriber video telephone service. Head-end or station
channels can be connected to subcenter terminals using floating
or fixed-matrix types of switching.

For large networks, the use of fixed-matrix video switching
is expensive for large-scale implementation, as well as present-
ing problems in size and volume in the switching centers, but
can be approached by optoelectronics, as previously described
and discussed further in Chap. 4. For small networks or mili-
tary applications, in which all terminals must have instant access
to all channels at all times, fixed-matrix switching is preferable.
When systems are to be designed and implemented in a large-city,
city-to-city, national, or international scale, the designs and
methodology for the switching must be very carefully planned.
Plans include taking into consideration present and future switching techniques and components, including optical, electrical, electrooptic, and optooptic devices and computer hardware and software.

Optical wired communities contain a large number of relatively
simple components. The complexity of these large networks comes
from the number of components, particularly the intricacy of their
interconnections, rather than from any great complexity of the individual components themselves. The predominant engineering
assumption of the late 1970s through the mid-1980s has been that
every component in a complex multiterminal system is there for a
reason. The assumption is based on the belief that the removal
or malfunction of any single component would cause the system,
or some substantial portion, to malfunction. Indeed, for many
systems based on point-to-point links as the central building
block for the large network, these assumptions are true. To
achieve reasonable economics for large optical networks, it is
possible to remove components, undergo a reorganization of
components and interconnections, and provide an efficient working system with many fewer components, at a substantially lower
cost per subscriber terminal.

It is possible using existing technology and components to
greatly reduce the total number of components required in an
optical wired community and still perform the required task. By
combining optical and electrical domains and new communications
techniques, it is possible to reduce some types of components by
more than 85%. These intriguing possibilities have created what

may be called electrooptical complexity theory (EOCT). The objectives and basis of the mathematics in EOCT are to determine the minimum number of components needed for large electrooptic networks. EOCT seeks to find the minimum number of components needed for given systems by developing new designs that call for fewer components, and by showing that a certain number of components will be needed no matter what design is followed. Innovative designs can greatly reduce system costs, both for the initial installation and subsequent maintenance and expansions.

Complexity theory is strictly concerned with the number of components in a system. The word *complexity* to most people implies the sophistication of components and the intricacy of their interconnection. In EOCT, systems with fewer components are obtained by interconnecting electrical and optical components in a more intricate way. Electrooptic complexity theory is at present more difficult to comprehend than the more traditional complexity theory, which usually deals with the electrical domain or a single transmission medium. The advent of new electrooptic components and devices permits what may also be called transparency. This optical phenomena may be defined as any case in which one wavelength or group of wavelengths can freely pass a switching point or mode.

This speed-of-light passage or transparency using different wavelengths does not occur in the electrical domain. Frequency transparency is possible in both optical and electrical domains. In the simplest case of two signals originating from point A, each has a separate emission wavelength. At point B, perhaps a switching center, the signal is transparent to a simple optical coating on a fiber or lens and passes freely to point C. The other wavelength is deflected optically or electrooptically for signal processing. The wide-bandwidth, low-attenuation properties of fibers makes this type of transparency in optical communications a very attractive tool for component reduction.

The objective of a switching center in an optical network is to provide pathways or connections for calls or requests from subscribers to the local head end or to distant centers. During operation, many requests are in progress at the same time, and the switching system must thus provide many simultaneous pathways with a waiting time acceptable to the user. Waiting or queueing times for different subcenter services or functions often vary. For example, connection by CATV channels is expected to be fast, but connection to specialized data banks may understandably be designed for a longer waiting period. Actual acceptable delays are a function of culture and user mentality and vary within a community and particularly among different nations and cultures. For example, a waiting time of 10 to 15 sec may be acceptable in France, but in the United States and Japan queue times of 2 to 5 sec are considered acceptable by users.

In a wired community with many subscriber terminals, at any given moment in time a certain number of user requests are in progress, each involving different switches for each pathway, so that transmissions over each of the paths do not interfere. As subscribers change channels or activate or deactivate their terminals, a new pathway must be established without disturbing any of the other subscribers. The most simplistic approach to the construction of a switching network is to provide a separate switch for each possible channel request. When a request arrives on a particular trunk, destined for a particular channel or other subscriber, the appropriate switch is closed. When the request is terminated, the switch is opened. A network designed in this way is called a crossbar because it is often implemented as a set of horizontal bars corresponding to trunks, crossed with a set of vertical bars corresponding to subcenters, with a switch located at each point of intersection.

To determine the efficiency of the crossbar design, consider a network capable of handling N requests, one with N trunks and N subscribers. A total of N X N, or N^2, different requests can be made through the network. Thus, there are N^2 switches in a crossbar. The number of switches required for a similar network to handle 4N requests is not four times the number required for a network that can handle N requests, but 4^2, or 16 times that number. This phenomenon of disproportionate growth is called *diseconomy of scale*. This diseconomy of scale reflects remarkably large differences in the required number of switches in a network, when relatively small systems like Hi-OVIS, with a few hundred subscribers, is compared to a large system with thousands of subscribers. In real networks, the number of trunks is not equal to the number of subscribers but is usually proportional. Also, in a real switching system, the cost is not only judged by the number of switches but also by the interconnecting wires, frames, and other hardware and required software.

A design calling for N^2 switches in a network that handles no more than N requests at a time is clearly not the best. Many other designs can be used to improve upon the required number of switches in the wired community network. A sparse crossbar, for example, is a crossbar switching network from which many switches have been removed but that fulfills the requirement that every group consisting of a third of the horizontal bars is connected by switches to more than two-thirds of the vertical bars. By interconnecting sparse crossbars, it is possible to construct switching networks with no more than cN log N switches, where c is a constant.

Networks that provide a path for any request that arrives on a free trunk, destined for a free (unengaged) subscriber, are called nonblocking networks. Other switching network configurations include seldom-blocking networks, which provide paths for most calls but not all. In a wired community of the 1980s, it is not necessary to guarantee a path for every call in every possible state of the switching network, at all times. The state of a large network is often assumed to be random. However, certain subscribers form patterns for requests, by selecting some channels with high frequency and others perhaps not at all. By computer monitoring the switching requests over a period of time, switching centers using seldom-blocking networks can be rewired, or reappointed, to reduce the incidence of blocked requests and the time required for queueing (waiting).

Seldom-blocking networks have an unavoidable diseconomy of scale of the same type as nonblocking networks, in that the number of switches per call grows as the logarithm of the number of calls. At present in the electrical switching domain, it appears that seldom-blocking networks are both easier and cheaper to construct than are nonblocking ones. Figure 3.13 shows the number of requests (subscribers) versus the number of required switches per request for nonblocking and seldom-blocking networks using switching in the electrical and optical domains.

In fiber optics and electrooptics, the complexity of switching analyses has the added dimension of light or optics, in addition to electronics. The most common comprehension of this difference in fiber optics is the use of wavelength-division multiplexing (WDM). In WDM, two or more wavelengths can transmit on the same fiber at the same time, each operating independently of the other, each of which may contain signals multiplexed in the frequency domain as well as in optical wavelengths. In practice, only 10 wavelengths, or 10 WDM, have been demonstrated to date, using lasers for each individual channel. Lasers have narrow spectral widths, usually less than 2 nm. New single-mode lasers, however, have been constructed with apparent

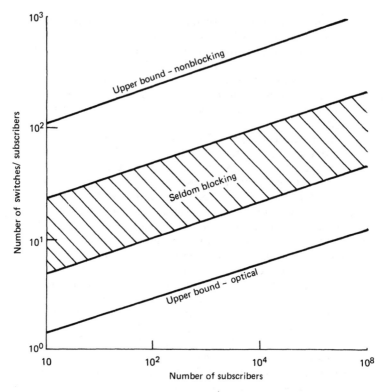

Figure 3.13. Switches required for networks versus subscribers.

spectral widths of less than 0.1 nm. With only existing fiber optics technology, using WDM requires optical channel spacing of
about 30 to 40 nm to maintain low cross talk and high signal quality using digital transmission. Given that the available wavelength
region is currently from 800 to 1600 nm, the number of available
optical channels per fiber is about 20 to 25.

The application of optics to complexity analysis permits en-
tirely new kinds of switching networks to be created, as well as
the implementation of new components to perform switching functions. In the electrical domain, switching is thought of as a state
of a component being on or off, or open or closed. In the electrooptic domain, switching is thought of as a state of component
being transmissive or reflective, active or passive, in addition
to being on or off, open or closed. In electrical digital switching,

components use transistors and gates and memory core to switch in various modes, such as space and time domains. In the electrooptic domain, switches use optical filters, opto-optic and electro-optic devices in addition to those used only in the electrical domains.

To determine the minimum number of switches required for a wired community communications network, various assumptions must be made as to the required channel capacity and online services. For the purpose of showing practical applications of complexity theory, a network similar to Hi-OVIS will be selected as the model community network. Thus, CATV channels, upstream data transmission, and the option for videophone service will be used in the system model. Note that, for all these wired city networks, the heaviest load on switching capacity is for videophone service. For the nonblocking network case, this service could require as many as N^2 (N = number of subscribers) switches, whereas CATV channel service and special request channels require many fewer switches.

For a practical nonblocking wired community switching network using only the electrical domain, the required number of switches is $49 \log_3 N$. For a seldom-blocking electrical switching system, the minimum number of switches is given by $3 \log_3 N$. For a practical system using both optics and electronics for switching functions, the minimum switching requirement is $\log_3 N$. Table 3.4 shows the three cases mentioned above as a function of the number of subscribers in the system.

To demonstrate the dramatic differences between these switching formats, take Hi-OVIS as an example of a small wired community. Hi-OVIS has 168 subscribers, each with a 30 X 2 video switch, and with the possibility of videophone service, a 168 X 168 switching matrix, using the nonblocking technique. The actual number of switches per subscriber is thus (30 X 2) + 168 = 228. The total number of switches is given by

(30 X 2 X 168) + (168 X 168) = 38,304

The value predicted in Table 3.4 for a nonblocking network is $49 \log_3 N = 49 \log_3 (168) = 228$, which is exactly that required for Hi-OVIS. For the practical optical switching case, the predicted value is $\log N = 5$. The total number of switches is given by 5 X 168 = 840.

The difference in the total required number of switches is staggering, even for such a small number of subscribers. As the number of subscribers increases, the difference becomes more

Table 3.4 Wired City Switching Requirements[a]

N	Practical nonblocking electrical: $49 \log_3 N$	Seldom-blocking electrical: $3 \log_3 N$	Practical optical switching: $\log_3 N$
10^2	205	13	4
10^3	308	19	6
10^4	410	25	8
10^5	513	32	10
10^6	616	38	13
10^7	719	44	15
10^8	821	51	17

[a]Relative ratios of switching requirements: nonblocking/seldom-blocking = 16:1; nonblocking/optical = 49:1; seldom-blocking = 3:1.

pronounced. The relative difference between a practical optical network and a nonblocking electrical type is 49:1 per subscriber. The calculation for the total number of switches quickly shows the large difference in costs for these various types of switching networks. Figure 3.13 shows the upper and lower bounds for the various network types plotted as the required number of switches per subscriber as a function of the total number of subscribers. The elimination of videophone service would greatly diminish the switching requirements for all network types.

To comprehend the large difference in savings by using seldom-blocking and practical optical networks, an arbitrary value of about $10 can be assigned for each installed simple switch in the wired city networks. This complete installed cost includes all wiring, connections, equipment racks, etc., to give a rough idea of the expenses involved in switching requirements. For the case of Hi-OVIS, the total switching cost would be $383,040. For the optical case, the total cost would be $8400. For a large city, assume $N = 10^6$. Then, the switching costs for the nonblocking case would be $6,160,000,000, for the seldom-blocking network $380,000,000, and the optical $130,000,000.

In the practical world over the next several decades we believe we will see a migration of optical transmission technology into the telecommunications networks, and optical switching will begin to play an increasingly important role.

REFERENCES

1. E. A. Harrington, Issues in Optical Switching Network Design, *Fiber Optic Communications, Information Gatekeepers, Inc.*, Chicago, 1978, pp. 17-20.
2. H. Kogelnik, Review of Integrated Optics, *Fiber and Integrated Optics 1*, No. 3, 1978.
3. V. E. Benes, *Mathematical Theory of Connecting Networks*, Academic Press, New York, 1965.
4. R. A. Steinberg and T. G. Giallorenzi, Performance Limitations Imposed on Optical Waveguide Switches and Modulators, Polarization, Applied Optics *15*, No. 10, October 1976, 2440-2453.
5. M. Inbar, Optical Channels in Distributed Processing, *Fiber Optics for Communication and Control*, Vol. 224, SPIE, 1980, pp. 7-62.
6. J. H. Monahan, Opportunities and Challenges in the Evolution of Local Area Networks, *Proceedings of the Local Area Communications Network Symposium, Boston, May 1979*.
7. M. Inbar, A Fiber Optic Based High Data Rate Computer Communication System, *Conference on Laser and Electro-Optical Systems, San Diego, February 1980*.
8. E. G. Rawson and R. M. Melcalfe, Fibernet, Multimode Optical Fibers for Local Loop Computer Networks, IEE Transactions on Communication *COM-26*, No. 7, July 1978, 983-990.
9. E. G. Rawson, Applications of Fiber Optics to Local Networks, *Proceedings of the Local Area Communications Network Symposium, Boston, May 1979*, pp. 155-168.
10. G. Hanke, Modulation of GaAs LED's in the GBit/sec Range, Proceedings of the ICC *3*, No. 44.3.1, June 1979.
11. D. Y. Farber, A Ring Network, *Datamation 21*, No. 2, February 1975, 44-46.
12. S. Blauman, Labeled Slot Multiplexing, *Proceedings of the Lowell Berkeley Conference on Distributed Data Management and Computer Networks, Lowell, Mass., August 1978*, pp. 309-322.

4

Traffic Control, Nonblocking Networks, and Local Area Networks

4.1 INTRODUCTION

Until about the mid-1970s, switching networks and traffic concepts followed a traditional evolution based largely on metallic wire conductor systems. As optical waveguide communications began growing at an increasing rate at about that time, an evolution based largely on fiber optics (FO) began to take hold. Its anticipated implementation is indicated in Fig. 4.1, with anticipated specialized optoelectric components of the mid-1980s and early 1990s shown in Table 4.1. Figure 4.2 shows the anticipated growth of component characteristics.

The traffic control of the future is being "wedded" to the in-place traffic control of the 1980s, and the present characteristics of coordinate switching networks is described in the following section.

In the final section we deal with the evolution of distributed optoelectric switching, switching centers, and the emerging optoelectronic PABX and local area networks (LAN) anticipated for the mid-1980s and through the balance of the twentieth century.

4.2 TRAFFIC CONTROL IN TELECOMMUNICATIONS

The definition of terms currently used in coordinate switching networks is given in the Glossary. A coordinate switch itself is used in switching networks to interconnect a number of inlet terminals to a number of outlet terminals. It is customary that every inlet and outlet be associated with a row or column where connecting devices are provided at the intersections. These are called *crosspoints*.

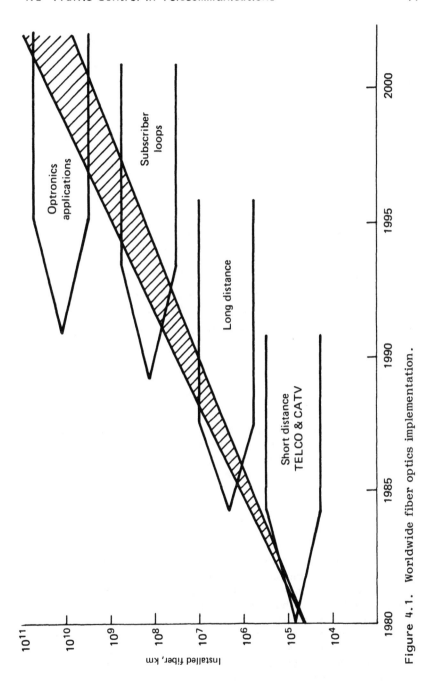

Figure 4.1. Worldwide fiber optics implementation.

Table 4.1 Anticipated Specialized Optoelectronic Components of the 1980s through 1990s

---

Optical LSI CODEC for telephone installations

Optical VLSI units for video (A/D) and (D/A) converters

Broad-band multiple-wavelength switching for EO multiterminal integrated networks and systems

EO high-speed telephone switching equipment for one or multiple wavelength transmission systems

Mass production of keyboards and terminals for interactive CATV and other two-way networks

Mass production of WDM components for 2 to 10 wavelengths with insertion losses <2 dB per channel and cross talk <40 dB

Mass production of single-mode fiber connectors and splices

Electrooptic tunable filters coupled to microprocessors for network switching and computing and wavelength multiplexing

Optical computers for high-speed analog, digital, and hybrid systems

Trigger laser amplifiers for high-bit-rate receivers and repeaters

---

Typical electromechanical coordinate switches of the 1980s have individual relays bringing about the contact, and the number of complete coil and contact relays becomes the product of the number of inlets and outlets. Another alternative has been to provide the entire crosspoint array with a crossbar switch in which a single relay coil is related to each row and column of the switch. By merging row and column coils, individual sets of crosspoint contacts can be closed.

In a more generic sense, the connecting device may be any two-state device that exhibits a low-impedance and a high-impedance state, for example, an optoelectronic device.

If the number of inlets and/or outlets becomes large, the number of crosspoints can be reduced appreciably. This can be done by replacing a single-coordinate switch by a number of interconnected smaller coordinate switches. This was described in Chap. 1 by arranging a multistage network in a number of ways. This will be further described by considering the properties of single-stage versus multistage coordinate switching networks.

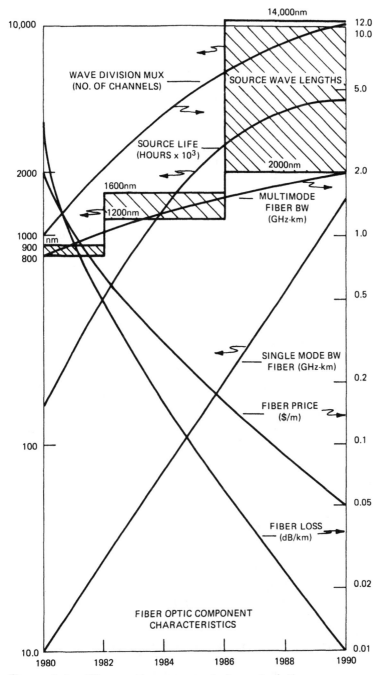

Figure 4.2. Fiber optic component characteristics.

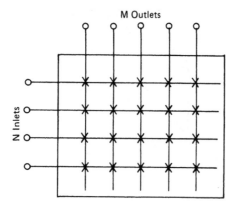

Figure 4.3. Single-stage square switching matrix with full availability (x = crosspoint contacts).

4.2.1 Single-Stage Coordinate Switches

Figure 4.3 shows a full-availability single-stage square coordinate
switch or switching matrix. With it, N line inlets are connected to
M line outlets. For full availability, interconnection is possible at
every crosspoint and the number of crosspoints available is NM,
providing a nonblocking condition. By limiting availability, an
economy of crosspoint contacts is obtained and a limited-availability switching matrix can be achieved.

Figure 4.4 shows a single-stage square switching matrix with
limited-availability interconnection possible, since every inlet has
access to only three outlets. This is a homogeneous grading, as
each set of three outlets is unique.

Figure 4.5 shows a single-stage folded triangular switching
matrix where the outlets and inlets are connected to the same
lines as the inlets; it is therefore nonblocking, with the number
of crosspoint contacts $N(N - 1)/2$.

4.2.2 Multistage Coordinate Switching Networks

Reference data handbooks, such as *Reference Data for Radio Engineers*, typically show details of configuration of multistage networks, such as for three and more stage networks. In this book
we illustrate some of the considerations of a three-stage network,
as it sets down the basic principles. We also restrict the number
of inlets and outlets for the same reason (although handbooks typically go to cases of 100 inlets and outlets).

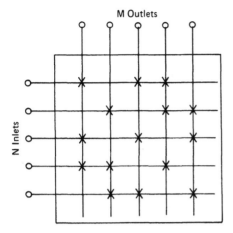

Figure 4.4. Single-stage square switching matrix with limited availability.

In the general case of a three-stage network with N/M primary switches with M outlets each, to interconnect a total of N inlets to M outlets, we illustrate these cases in Figs. 4.6 through 4.10. For extensions to larger stage networks, typically up to eight in current large-scale practice, we refer you to extrapolations from standard data handbooks designed for electrical and electronic and radio considerations.

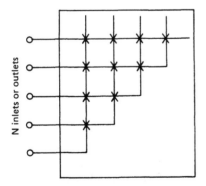

Figure 4.5. Single-stage folded triangular switching matrix with full availability.

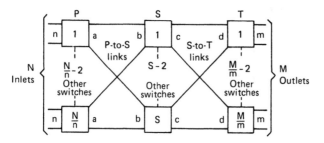

Figure 4.6. Three-stage network, general case. (Reproduced with permission, *Reference Data for Radio Engineers*, Sixth Edition, Third Printing, 1979, Howard W. Sams & Co., Inc., Indianapolis, Indiana.)

Referring to Fig. 4.6, the requirements for single-linkage spread connections between stages are

$$a = d = S \qquad b = \frac{N}{n} \qquad c = \frac{M}{m}$$

The number of crosspoints is given as

$$X = S\left(N + M + \frac{NM}{nm}\right)$$

The degree of internal blocking is determined by the relationship of secondary switches to the number of inlets and outlets.

Figure 4.7 illustrates the symmetrical case, where part of the network is illustrated, where $N = M$ and $n = m$. The expression for the number of crosspoints then simplifies to

$$X = S\left(2N + \frac{N^2}{n^2}\right)$$

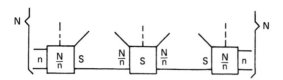

Figure 4.7. Symmetrical three-stage network. (Reproduced with permission, *Reference Data for Radio Engineers*, Sixth Edition, Third Printing, 1979, Howard W. Sams & Co., Inc., Indianapolis, Indiana.)

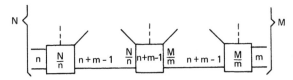

Figure 4.8. Nonblocking three-stage network. (Reproduced with permission, *Reference Data for Radio Engineers*, Sixth Edition, Third Printing, 1979, Howard W. Sams & Co., Inc., Indianapolis, Indiana.)

Figure 4.8 shows the general case for nonblocking three-stage networks. When N is not equal to M, the nonblocking condition is given by

$$S = n + m - 1$$

and the number of required crosspoints is

$$X = (n + m - 1) \left(N + M + \frac{NM}{nm} \right)$$

When m = n, a minimum number of crosspoints is obtained, and when n satisfies the equation

$$\frac{NM}{N + M} = \frac{n^3}{n - 1}$$

The unit of traffic intensity is a pure number, as it is a ratio without dimension. One erlang of traffic intensity in one traffic path means continuous occupancy of that path, irrespective of the duration of path occupancy. The erlang was adopted in 1946 by the CCIF (now the CCITT) as the international unit of traffic intensity. This definition requires an expansion in the near future as optical waveguides and switching begin to play a major

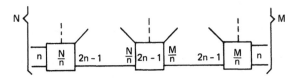

Figure 4.9. Nonblocking three-stage network with minimum crosspoints. (Reproduced with permission, *Reference Data for Radio Engineers*, Sixth Edition, Third Printing, 1979, Howard W. Sams & Co., Inc., Indianapolis, Indiana.)

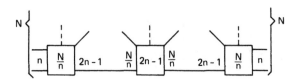

Figure 4.10. Symmetrical nonblocking network with minimum cross-points. (Reproduced with permission, *Reference Data for Radio Engineers*, Sixth Edition, Third Printing, 1979, Howard W. Sams & Co., Inc., Indianapolis, Indiana.)

role in traffic. For example, with wave-division multiplexing, many (or more than one) signals may occupy a path for different lengths of time. One will possibly have to also define an "optical erlang."

4.3 EVOLUTION OF DISTRIBUTED OPTOELECTRONIC SWITCH-ING, SWITCHING CENTERS, OPTOELECTRONIC PABX, AND LOCAL AREA NETWORKS

Although it is evident that optical switching systems in the mid-1980s can directly exchange an optical signal without electrooptical (E/O) and optoelectronic (O/E) conversion, such as for an optical PABX or optical toll/central office, and optical local area networks, this book illustrates the evolutionary step currently in experimental use in places like Japan and Canada of optoelectronic switching and optoelectronic PABX. We also cite the rapidly developing area of optical local area networks.

4.3.1 Role of Fiber Optics in Local Area Networks*

In the United States, since AT&T's divestiture from its local operating companies, local distribution has been placed at the center of the reorganization of the telecommunications industry. In the domain of bringing sophisticated computer power and telecommunications services to everyone's home and office, fierce competition is expected. The new legislation will speed up the introduction of the innovative local distribution networks needed to establish the information society. Fiber optics is perceived by the

---

*This subsection is a reedited version of material supplied with the permission of Dr. Edouard Y. Rocher of Aetna Telecommunications Consultants, Centerville, Massachusetts, 1983.

telecommunication industry, as well as by the end user, as having
a key role. The purpose of this section is to discuss the why,
when, and how of the industry.

Present Environment: Telephone, CATV, and LAN Products

Local distribution networks can be defined as the facilities to
which the customer connects telecommunication equipment for
transporting (transmitting and switching) information (voice,
data, or video) from one type of equipment to another. Besides
the kind of transport services that they provide, local networks
are also characterized by the size of the area they cover, i.e.,
what is meant by local (building, plant, campus, or metropoli-
tan) and whether they are private (one customer account) or
public (a multiplicity of customer accounts).

The market segment addressed by a local distribution network
has several important impacts on the dynamic evolution of the size
of the market and on the characteristics of the distribution system itself: number of connection points (from ten or less, to tens
of thousands), distance (from a few meters, to as much as 80 km),
and size and characteristics of the switch. Finally, local area networks can be put into three categories according to the basic information transport services they are optimized for: voice (telephone network), video (CATV networks), or data (local area
networks).

For each of the three categories, a wide variety of local distribution systems is available on the market. In each instance,
FO technology is starting to displace conventional transmission
technologies, but only inside the network, and not yet at the
user equipment (distribution) level, e.g., digital FT3 carriers
for the telephone network, main trunks for CATV network, and
remote segments interconnection for Ethernet LAN. Fiber optics,
and the services it will provide, cannot effectively compete with
the implemented plants at the distribution level until such time
that it becomes cost effective and the potential multimillion FO
component units develop.

With close to 100 million subscriber lines in the United States,
the telephone network is, by far, the most ubiquitous local distribution network. Although over the last 25 years digital techniques have been introduced inside the network at a rapid pace,
first for control and transport and then for switching, almost all
local distribution (local loop and associated switch) is still analog.
Figure 4.11 shows the hierarchical structure of the telephone network. The local plant (end offices and station loop) represents
the major investment, an estimated $80 billion, of the telephone

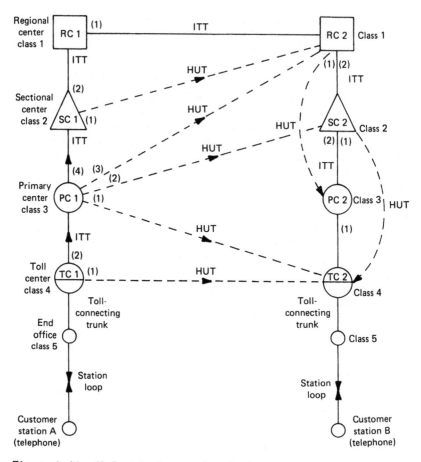

Figure 4.11. U.S. telephone network structure.

industry. At the present time, FO technology has found industrial applications for digitalized voice transmission on long-haul
links and on submarine cables (TCT, HUT, and ITT in Fig. 4.11).

With close to 20 million subscribers in the United States, CATV
(community antenna television, or cable television) networks are
the second major local information distribution systems. They are
optimized for broadcasting several 6 MHz TV channels and FM
audio programs multiplexed in a broad-band signal (300 MHz)
from a central point, the head end, to every user. A few systems
provide some bandwidth to return information from the user to the

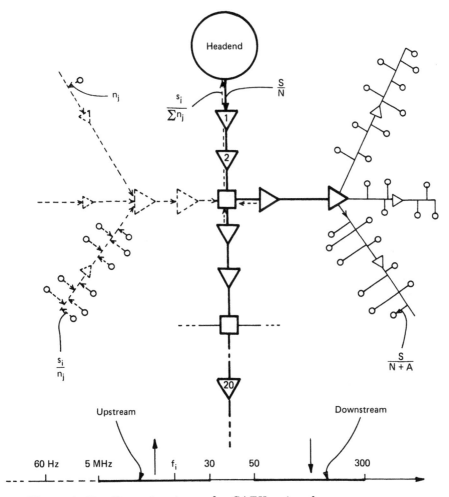

Figure 4.12. Tree structure of a CATV network.

head end, but due to the tree structure of the network, the number of entry points must be limited (Fig. 4.12).

Local area networks are new distribution networks optimized
for data communications. Despite the fact that there are more
than 40 different LAN offerings at present, the total number of
devices actually attached to LAN is still far from reaching a million. Table 4.2 gives a cross section of available LAN products.
The main purpose of the table is to show the wide variety of

Table 4.2 Typical LAN Offerings

Company, LAN name	Data	Voice	Video	Trans-mission speed (b/sec)	Broad band	Base band
Amdax Corp., Cablenet	Yes	No	Yes	7/14M	X	
Data General Corp.	Yes	No	No	2M		X
Datapoint Corp., Arcnet	Yes	No	No	2.5M		X
The Destek Group, Desnet	Yes	No	No	2M	X	X
Digital Microsystems, Hinet	Yes	No	No	500K		X
Interactive Systems/ 3M, Inc., Videodata	Yes	Yes	Yes	300K to 2.1M	X	
Network Systems Corp., Hyper-channel	Yes	No	No	50M		X
Proteon Associates, Inc., Pronet	Yes	No	No	10M		X
Sytek, Inc., Local-net Model 20/40	Yes	No	No	128K 2M	X	
Ungermann-Bass, Inc., Net/One	Yes	No/ Yes	No	10M /5M	X	/X
Wang Laboratories, Inc., Wangnet	Yes	Yes	Yes	9.6K to 12M	X	
Xerox Corp., Ethernet	Yes	No	No	10M		X

Topology	Cable medium	No. of sup. dev.	Geographic scan (km)	Price/dev. attachment	Installed base
Tree	Coaxial	16,000	80	$1400 w/60 dev.	12
Bus	Coaxial	32	1.6	$3400	0
Bus	Coaxial	255	6	$500	2000
Bus	Coaxial, twisted pair	4600	6	$265 w/8 dev.	0
Bus	Dual, twisted pair	255	0.3	$1500	1000
Bus	Coaxial	10,000	80	$400-$600	∿250
Bus	Coaxial	64	2	$40,000	902
Star/ring	Twisted pair, coaxial/FO	255	25	$600 w/16 dev.	10
Tree	Coaxial	20,000-60,000	80	$525 w/8 dev.	65
Bus	Coaxial	1024	2.5-16	$450 w/24 dev.	100/0
Tree	Coaxial	16,000	3	$360 w/10 dev.	0
Bus	Coaxial (FO)	1024	2.5	$250-$1000 dep. on devices	No. not released

objectives and technologies that characterize the burgeoning LAN industry. Generally, LAN are based on coaxial cable technology; they provided megabits per second at a price of $500 per device in 1983; they do not support full-function voice communications; and when based on broad band, they are adapted to the CATV technology and provide some video services. Most of the 40 LAN offerings are units that adapt the LAN connection standards to standard telecommunication interfaces (e.g., RS-232C) and standard services, e.g., 9.6 kb/sec. A major trend in the industry of 1983 is to comply with the Ethernet standard (10 Mb/sec baseband, CSMA/CD), which seems to be a de facto standard at this time, once the VLSI chips are available to run the Ethernet protocols. At least 70 equipment manufacturers have indicated that they intend to adopt the standard proposed by Xerox, Intel, and DEC in 1980.

LAN Market Trends: Impact on Fiber

The LAN market is driven by the need to more efficiently interconnect all the pieces of equipment that are based on micro- or miniprocessors. This market in turn is opened up by the rapid evolution of the very large scale integration (VLSI) technology (Fig. 4.13), making it possible to implement a large amount of processing power and of memory within each terminal device, at very low cost. The evolution of the terminal market is shown in Fig. 4.14 where it is compared with the evolution of the telephone and the television set markets. It can be expected that by the year 2000 there will be over 100 million data terminals in the United States. Figure 4.15 shows the same curves with a linear vertical scale; it shows clearly where the competition is going to be.

It is important to realize the differences among the three categories of market. With the telephone, an interactive exchange of information occurs in a transparent fashion between two individuals who are not part of the network; the intelligence is outside the network. With television, information is broadcast from one location to a multiplicity of viewers; there is some intelligence in the network, and an important part of the business is to generate the programs, etc. With data communications, an interactive exchange of information occurs between two intelligent pieces of equipment that are part of the network. The network can be subdivided into two complimentary parts: the distribution network, which performs the data transport function, and the systems network, which performs the information processing functions. One provides the physical link, the other the logical link.

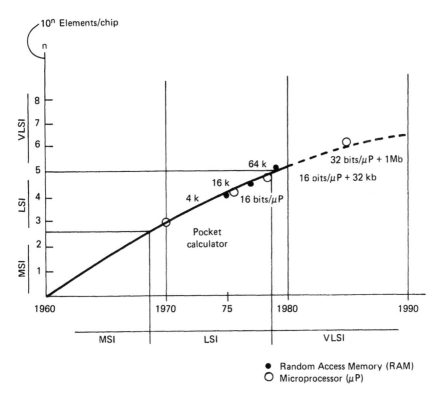

Figure 4.13. Maximum density of active components integrated on a silicon chip, 1970-1980.

The revenues of the data processing industry in the United States were projected to grow from $43 billion in 1980, to $80 billion in 1985, and to $150 billion in 1990, with an increasing contribution from interactive applications and thus an increasing dependency on data communications.

The bulk of data communications is still provided by the telephone network, and the appearance of the first LAN represents a new trend to bypass the telephone network for local distribution. With the evolution of these technologies, it is clear that it has become possible to offer, at very low cost, much wider bandwidth than the conventional 1.2 to 9.6 kb/sec capability of the analog transparent telephone network. In fact, what the customer is looking for is the information outlet, the equivalent of the standard power outlet (Fig. 4.16). If properly designed,

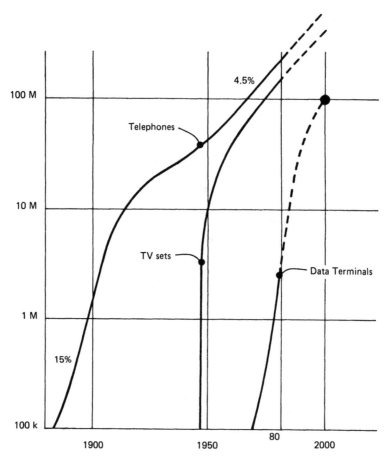

Figure 4.14. Number of telephones, TV sets, and terminals in the United States, 1880-2000.

the information jack in the wall and the associated stable wiring will lead to savings in local network maintenance and in greater utilization flexibility by increasing terminal mobility and by allowing easy upgrading. The first requirement for the information outlet is wide band. It should be possible to plug into the same outlet a wide variety of devices, e.g., sensors (bits per second), conventional terminals (kilobits per second, bursty), digital voice (kilobits per second, continuous), image, graphics

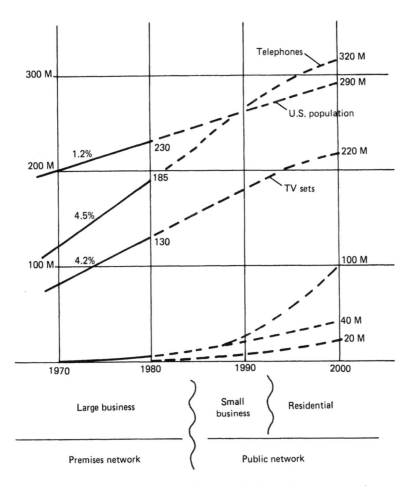

Figure 4.15. A scenario for data terminal market penetration.

(megabits per second, bursty), etc. Associated with the notion
of information outlet is the system notion of universal interface,
which raises architecture issues.

The coming decade will see a confrontation between two trends
in the United States: the conventional telecommunications industry will offer 64 kb/sec line switched as a wide-band service (digital private brand exchange, PBX, integrated system digital network, ISDN); the data processing industry will offer megabit per

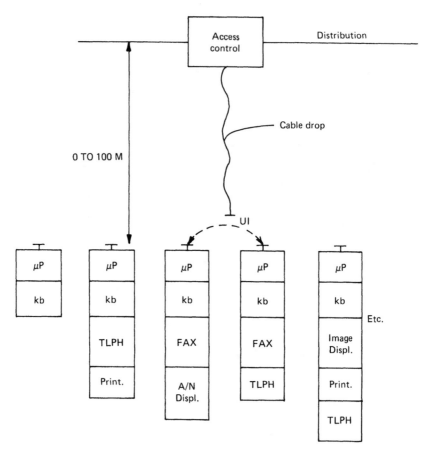

Figure 4.16. Universal interface LAN.

second block-switched services on LAN. LAN of the first generation do not support voice; the second generation will. One determining factor in the PBX-LAN confrontation will be the value of the applications requiring megabits per second capability that LAN will be able to support and PBX will not. At the present time, these applications are few, mainly because the industry has traditionally been limited by the narrow bandwidth of the telephone network and by the unavailability of low-cost intelligence. Nevertheless, applications can be developed for image, graphics, or facsimile that will require the transfer of several megabits in less than 1 sec for getting 1-sec response times. For example, by 1990, digital optical disks will store a 250-page book for about $0.10; systems will store the knowledge of one library (e.g., 60,000 books and 6000 periodicals) and provide a 10-sec book seek time and then a 150-msec page access time. With the progress in VLSI, there is an increasing trend toward distributed processing. As cost permits an increasing amount of memory in each device, one way to prove system utilization flexibility will be to load programs from a remote server and execute them locally. These types of applications will generate a very bursty traffic, high bandwidth (megabits per second) during short periods of time (milliseconds or seconds), in contrast with the relatively continuous traffic generated by digital voice or video. Most LAN take advantage of the bursty nature of the data traffic to share the transmission facility in an efficient way to reduce costs.

LAN Architecture: Role of Fiber

In order to organize the telecommunications systems, the industry has adopted an architecture model that defines seven layers of functions to be performed in a fixed sequence for two pieces of equipment to exchange information (Fig. 4.17). The architecture specifies the protocol for communications between peer layers; these protocols are product independent. Between layers, interfaces are defined that are all product dependent, except one — the interface between the network and the equipment (Fig. 4.16). The definition of the information outlet supposes definition of this interface and also of all the protocols associated with the layers or sublayers below the interface. Figure 4.18 shows a possible sublayering of the lower two layers. Ethernet's interface is defined above sublayer 2.4. The model introduces the notion of functional transparency. For example, above sublayer 1.2 the architecture is independent of the medium (twisted pair, coax, baseband, FO, etc.). It also shows that the most important characteristic of LAN is network topology, and more specifically whether the switching is centralized (star or tree) or decentralized (tree, bus, loop, and ring). Table 4.3 is a summary of the attributes generally associated with the different topologies.

FO technology will be particularly well adapted to star and ring topologies, star because the low losses permit full utilization of the bandwidth over long distances without repeaters, ring because each network interface unit acts as a distributed

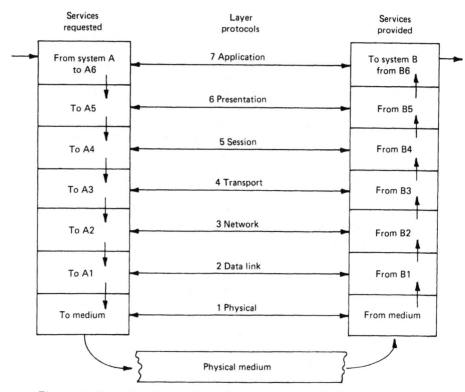

Figure 4.17. Abstract reference model, open system interconnection.

switch and as a repeater. The selection of the preferred topology depends upon the combination of voice, video, and data services to be provided by the distribution network, installation costs, and operating costs. The choice will have a deep impact on the specification of FO technologies for LAN (analog/digital, losses, bandwidth, cable length, connectors, types of cables, etc.).

4.3.2 Passive and Active Star Configurations for Office Communications Systems

This section gives one example of a centrally switched star configuration for office communications systems. Passive stars are also being used and will see growing use in the mid 1980s and beyond. They can be particularly attractive for mobile platform uses, and they are very sparing of energy.

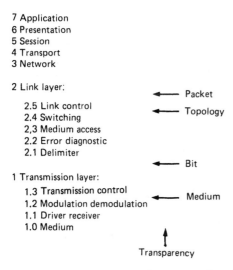

7 Application
6 Presentation
5 Session
4 Transport
3 Network

2 Link layer: ◄——— Packet

   2.5 Link control   ◄——— Topology
   2.4 Switching
   2,3 Medium access
   2.2 Error diagnostic
   2.1 Delimiter
                                 ◄——— Bit

1 Transmission layer:
   1.3 Transmission control
   1.2 Modulation demodulation   ◄——— Medium
   1.1 Driver receiver
   1.0 Medium
                              ↑
                        Transparency

Figure 4.18. Sublayering example: layers 1 and 2.

Table 4.3 LAN Topologies

Star: centralized
  Telco
  Voice-data → switched video
  Copper twisted pair → fiber
  Line switching

Tree: centralized and decentralized
  CATV
  Video broadcasting
  Coaxial - broad band
  FDM + packet switching or TDMA

Bus: decentralized
  DP
  Wide-band data
  New wiring; coaxial - baseband
  Packet switching + CSMA/CD

Ring: decentralized
  DP
  Wide-band data
  New wiring: Shielded twisted pair, coaxial fiber
  Packet switching + token

The centrally switched star configuration has been well described by Elmer H. Hara of the Department of Communications of Canada at *Globecom 1982 (Proceedings of IEEE Global Telecommunications Conference,* pages 961-965. He illustrated use of a broad-band analog PBX and time-division multiplexed amplitude (TDM-PAM) transmission using optical fibers. This permitted a network of many analog and digital channels. Prior experimental results using optoelectronic switches had shown that a PBX matrix switch of size 30 X 30 for 64-kb/sec digital signals or 100-kHz analog signals was possible. With such matrix switches, cross-talk loss and isolation (on/off power ratio) of more than 60 dB is readily achievable. A digital PBX or analog PBX may be used. The broadband analog PBX provides a LAN that is transparent to various transmission protocols and as such offers ready flexibility in the choice of office digital equipment. The use of optical fiber further permits office video services by virtue of its considerable transmission bandwidth.

4.3.3 Some General Comments on Switching and Interconnection in Local Area Networks

The second generation of LAN have bandwidth distance products of up to about 50 MdB/km. Data rates are often limited by higher level computer software protocols rather than by the transmission medium itself. Optical taps with negligible perturbation on the main network are under development, as are other network mode interfaces compatible with coaxial networks.

For local data communications bandwidth, the distance achievable at present is as high as 200 MHz/km.

Standardization is difficult in this new technology, as many competitive approaches are available to achieve similar end results. Functional and physical interfaces between the fiber transmission medium and the associated hardware not necessarily unique to fiber optics is now possible.

Point-to-point links typically use star or ring or loop networks in LAN. For these localized well-defined networks, optical power budgets are readily specified for the network, independent of tandem modes in various numbers. Distributed fiber optic loop networks have been implemented. Some of these have automatic bypass switching and soft failure network redundancy configurations. With the advent of low-perturbation bidirectional taps in optical data transmission, optical data signals can be coupled efficiently into and out of the main fiber network. It is now possible to avoid severing the main network fiber to introduce the coupler for a new node.

Bibliography

Akimaru, H., and Nishimura, T., The Derivatives of Erlang's B-formula, Review of the Electrical Communication Laboratory 11, 428-444, 1963.

Anderberg, M., Fried, T., and Rudberg, A., Optimization of Exchange Locations and Boundaries in Local Telephone Networks, Proceedings of the 7th International Teletraffic Congress, Stockholm, 1973, pp. 424/1-7.

Aro, E., Stored Program Control-assisted Electromechanical Switching — An Overview, Proc. IEEE No. 9, September 1977.

Atkinson, J., *Telephony*, Sir Isaac Pitman and Sons, Ltd., London, 1957.

Atkinson, J., *Telephony*, Vol. 2, Sir Isaac Pitman and Sons, Ltd., London, 1959.

ATT Blue Book, Sec. 5, Par. 5, International Telecommunications Union, Geneva, 1975.

Backman, G., and Penn, R., Automatic Identified Outward Dialing for PBX's: Central Office Facilities, Bell Laboratories Record 46, 112-116, April 1968.

Batra, V., Grounding Electronic Switching Systems, Telephony May 1974.

Bazlen, D., Link Systems with Bothway Connections and Outgoing Finite Source Traffic, Proceedings of the 7th International Teletraffic Congress, Stockholm, 1973, pp. 316/1-8.

Bear, D., The Use of "Pure Chance" and "Smoothed" Traffic Tables in Telephone Engineering, Proceedings of the 2nd International Teletraffic Congress, The Hague, 1958.

Bear, D., Some Theories of Telephone Traffic Distribution: A Critical Survey, Proceedings of the 7th International Teletraffic Congress, Stockholm, 1973, pp. 531/1-5.

Bear, D., Principles of Telecommunications Traffic Engineering, IEEE Telecommunications Series 2, Peter Peregrinus, London, 1976.

Bear, D., and Seymour, C., A Traffic Prediction Model for a Telephone Exchange Network, Proceedings of the 7th International Teletraffic Congress, Stockholm, 1973, pp. 536/1-5.

Beckmann, P., An Introduction to Elementary Queuing Theory and Telephone Traffic, The Golem Press, Boulder, Colorado, 1968.

Benes, V., Mathematical Theory of Connecting Networks and Telephone Traffic, Academic Press, New York, 1965.

Benes, V., Traffic in Connecting Networks When Existing Calls Are Rearranged, Bell System Technical Journal 49, 1471-1482, 1970.

Berkeley, G., Traffic and Trunking Principles in Automatic Telephony, 2nd revised edition, Ernest Benn, Ltd., London, 1949.

Berry, L., An Explicit Formula for Dimensioning Links Offered Overflow Traffic, ATR 8, 13-17, 1978.

Biddalph, R., Budlong, A., Casterline, R., Funk, D., and Goeller, L., Lines, Trunk Junctor and Service Circuits for No. 1 EES, The Bell System Technical Journal 43, 2323, September 1964.

Bininda, N., and Daisenberger, G., Recursive and Iterative Formulae for the Calculation of Losses in Link Systems of any Description, 5th International Teletraffic Congress, New York, 1967, pp. 318-326.

Bocqueho, P., and Triboulet, M., Technical and Cost Considerations on Optical and Electrical Multiplexing for Fiber Optic Systems, Fiber Optics and Communications Proceedings, Information Gatekeepers, Inc., Brookline, Massachusetts, 1981, pp. 7-12.

Boehlen, C., Telephone Traffic Definitions, Telephone Engineer and Management para. 2b, 2g, February-March 1944.

Bowers, T., Blocking in 3-stage "Folded" Switching Arrays, IEEE Paper No. CP 63-1461, 1967.

Bowers, T., Derivation of Blocking Formulae for 3-stage "Folded" Switching Arrays, IEEE Paper No. CP 63-1462, 1967.

Bretschneider, G., Exact Loss Calculations of Gradings, 5th International Teletraffic Congress, New York, 1967, pp. 162-169.

Bridgford, J., The Geometric Group Concept and Its Application to the Dimensioning of Link Access Systems, Proceedings of the 4th International Teletraffic Congress, London, 1964, paper 13.

Briley, B., and Toy, W., Telecommunications Processors, Proceedings of the IEEE 65, No. 9, September 1977.

Brockmeyer, E., The Simple Overflow Problem in the Theory of Telephone Traffic, Teleteknik 5, 361-374, 1954.

Brockmeyer, E., Halstom, H., and Jensen, A., *Life and Works of A. K. Erlang*, Transactions of the Danish Academy of Technical Sciences, Copenhagen, 2, 21, 1948.

Brockmeyer, E., et al., *The Life and Works of A. K. Erlang*, Acta Polytechnica Scandinavia, The Danish Academy of Technical Sciences, Copenhagen, 1960.

Buchner, M., and Neal, S., Inherent Load Balancing in Step-by-Step Switching Systems, Bell System Technical Journal 50, 135-165, 1971.

Burke, P., The Output of a Queuing System, Journal of Operations Research Society of America 4, 699-704, 1956.

Burke, P., Equilibrium Delay Distribution for One Channel with Constant Holding Time, Poisson Input and Random Service, Bell System Technical Journal 38, 1021-1031, 1959.

Casey, J., and Shimasaki, N., Optimal Dimensioning of a Satellite Network Using Alternate Routing Concepts, 6th International Teletraffic Congress, Munich, 1970, pp. 344/1-8.

CCITT Blue Book, Measurements of Traffic Flow, Recommendation E90, Vol. II, 1964, p. 143.

CCITT Blue Book, Determination of the Number of Circuits Necessary to Carry a Given Traffic Flow, Recommendation E95, Vol. II, 1964, p. 203.

CCITT Blue Book, Geneva, Vol. II, 1965, pp. 70, 239.

CCITT Green Book, International Telecommunications Union, Geneva, Vol. III, 1973, part 1, sect. 1.

CCITT Green Book, International Telecommunications Union, Geneva, Vol. III, 1973, (Q 2 D).

CCITT Green Book, International Telecommunications Union, Geneva, Vol. VI-1, 1973, pp. 25-40.

CCITT Green Book, International Telecommunications Union, Geneva, Vol. VI-1, 1973, Recommendation Q45.

CCITT Green Book, International Telecommunications Union, Geneva, Vol. VI(IX), 1973.

CCITT Green Book, International Telecommunications Union, Geneva, Vol. VI(X), 1973.

CCITT Green Book, International Telecommunications Union, Geneva, Vol. VI(XII), 1973.

CCITT Green Book, International Telecommunications Union, Geneva, Vol. VI(3), 1973, Chap. 3.

CCITT Green Book, International Telecommunications Union, Geneva, Vol. VI-XV, 1973.

CCITT Green Book, International Telecommunications Union, Geneva, Vol. XI-XVI, 1973.

CCITT Green Book, International Telecommunications Union, Geneva, Vol. VIII, 1973, Recommendation V3.

CCITT Orange Book, Q Recommendations (Recommendation Q), International Telecommunications Union, Geneva, 1977.

CCITT Red Book, International Telecommunications Union, Geneva, Vol. VI(5), 1960.

CCITT White Book, International Telecommunications Union, Geneva, Vol. 1, 1969.

Chan, W., and Chung, W., Waiting Time Distribution in Computer Controlled Queuing System, Proceedings of the IEEE *118*, No. 10, 1378-1382, 1971.

Chinnick, J., The Development of an Opto-Electronic Wideband Switching Array, Fiber Optics and Communications Proceedings, Information Gatekeepers, Inc., 1981, pp. 79-82.

Clos, C., A Study of Nonblocking Switching Methods, Bell System Technical Journal *32*, 406-424, 1953.

Cobham, A., Priority Assignment in Waiting Line Problems, Journal of Operations Research Society of America *2*, 70-76, 1954; *3*, 547, 1955.

Coffman, E., and Kleinrock, L., Some Feedback Queuing Models for Time-shared Systems, *Teletraffic Engineering Manual*, New York, 1967, pp. 288-304.

Cohen, J., Certain Delay Problems for a Full Availability Trunk Group Loaded by Two Traffic Sources, Communications News *16*, 105-115, 1956.

Coleman, R., Use of a Gate to Reduce the Variance of Delays in Queues with Random Service, Bell System Technical Journal *52*, 1403-1422, 1973.

A Course in Telephone Traffic Engineering, Australian Post Office, Planning Branch, Melbourne, 1967.

Cox, D., and Smith, W., *Queues*, Methuen, 1961.

Crider, G., and Foster, W., Automatic Telephone Dial Directory, IBM Technical Discipline Bulletin, December 1976.

Crommelin, C., Delay Probability Formulae, Post Office Electrical Engineering Journal *26*, 266-274, 1934.

Dahlbom, C., Common Channel Signaling — a New Flexible Interoffice Signaling Technique, International Switching Symposium, 1972, pp. 421-427.

Dartois, J., Lost Call Cleared Systems with Unbalanced Traffic Sources, Proceedings of the 6th International Teletraffic Congress, Munich, 1970, pp. 215/1-7.

Dartois, J., Metaconta L Medium Size Local Exchanges, Electrical Communications *48*, No. 3, 1973.

De Boer, J., Comparison of Random Selection and Selection with Fixed Starting Position in a Multi-Stage Network, Philips Telecommunications Review *31*, 148-155, 1973.

De Ferra, P., and Massetti, G., On the Quality of Telephone Service in an Automatic Network and Its Economical Expression, Proceedings of the 6th International Teletraffic Congress, Munich, 1970, pp. 135/1-6.

Descloux, A., *Delay Tables for Finite and Infinite-Source Systems*, McGraw-Hill, New York, 1962.

Dietrich, G., et al., *Teletraffic Engineering Manual*, Standard Electric Lorenz, Stuttgart, Germany, 1970.

Donaldson, J., Structured Programming, Datamation *52*, December 1973.

Duerdoth, W., and Seymour, C., A Quasi-Non-Blocking TDM Switch, Proceedings of the 7th International Teletraffic Congress, Stockholm, 1973, pp. 632/1-4.

Einarsson, K., Hakansson, L., Lindgren, E., and Tange, I., Simplified, Type of Gradings with Skipping, Teleteknik (English edition) 74-96, 1961.

Electrical Communication System Engineering Traffic, U.S. Department of the Army, TM-11-486-2, August 1956.

Elion, G., *Connectors, Splices and Couplers*, Information Gatekeepers, Inc., Brookline, Massachusetts, International Communications and Energy, Inc., 1978.

Elion, G., *Systems and Data Links*, Information Gatekeepers, Inc., Brookline, Massachusetts, International Communications and Energy, Inc., 1979.

Elion, G., and Elion, H., *Fiber Optics in Communications Systems*, Marcel Dekker, Inc., New York, 1978.

Elion, G., and Elion, H., *Electro-Optics Handbook*, Marcel Dekker, Inc., New York, 1979.

Elldin, A., Further Studies in Gradings with Random Hunting, Ericsson Tech. *13*, 177-257, 1957.

Elldin, A., On the Dependence Between the Two Stages in a Link System, Ericsson Tech. *2*, 185-259, 1961.

Erlang, A., Solution of Some Problems in the Theory of Probabilities of Significance in Automatic Telephone Exchanges, Post Office Electrical Engineering Journal *10*, 189-197, 1918.

Feiner, A., The Ferreed, The Bell System Technical Journal *43*, January 1964.

Feiner, A., and Hayward, W., No. 1 EES Switching Network Plan, The Bell System Technical Journal *43*, 2215, September 1964.

Fiber Optics Blueprints for the Future, International Communications and Energy, Inc. (client-limited edition), 1981.

Flood, J., *Telecommunications Networks*, IEE Telecommunications
    Series 1, Peter Peregrinus, London, 1975.

Flowers, T., Processors and Processing in Telephone Exchanges,
    Proceedings of the IEEE *119*, No. 3, March 1972.

Flowers, T., *Introduction to Exchange Systems*, John Wiley and
    Sons, New York, 1976.

Frank, H., and Chou, W., Topological Optimization of Computer
    Networks, Proceedings of the IEEE *60*, 1385-1397, 1972.

Frank, H., and Frisch, L., *Communication, Transmission and
    Transportation Networks*, Addison Wesley, Boston, 1971.

Freeman, R., *Telecommunications Transmission Handbook*, John
    Wiley and Sons, New York, 1975.

Freeman, R., *Telecommunication System Engineering*, John Wiley
    and Sons, New York, 1980.

Fultz, K., and Penick, D., The T1 Carrier System, Bell System
    Technical Journal 44, September 1965.

Functional Specification and Description Language (SDL), CCITT
    Orange Book, International Telecommunications Union, Geneva,
    Vol. V.1.4, 1977.

Fundamental Principles of Switching Circuits and Systems, Ameri-
    can Telephone and Telegraph Company, New York, 1963.

Gimpelson, L., Network Management: Design and Control of
    Communications Networks, Electrical Communications *49*,
    4-22, 1974.

Goeller, L., Design Background for Telephone Switching, *Lees
    ABC of the Telephone*, Geneva, Illinois, 1977.

Goleworth, H., Kyme, R., and Rowe, J., The Measurement of
    Telephone Traffic, Post Office Electrical Engineering Journal
    64, 227-233, 1972.

Gosztony, G., Full Availability One-Way and Both-Way Trunk
    Groups with Delay and Loss Type Traffic, Finite Number of
    Traffic Sources and Limited Queue Length, Proceedings of
    the 7th International Teletraffic Congress, Stockholm, 1973,
    pp. 341/1-8.

Gracey, A., Basic Theory Underlying Bell System Facilities Ca-
    pacity Tables, Transactions of the AIEE *69*, 238, 1950.

Grinsted, W., A Study of Telephone Traffic Problems with the
    Aid of the Principles of Probability, Post Office Electrical
    Engineering Journal *8*, 33-45, 1915.

Hara, E., MacDonald, R., and Tremblay, Y., Optoelectronic
    Switching with Avalanche Photodiodes, Technical Digest,
    Topical Meeting on Optical Fiber Communications, Washington,
    D.C., March 1979. IEEE catalog no. 79CH1431-6 QEA.

Hara, E., Machida, S., Ikeda, M., Kanbe, H., and Kimura, T., A High Speed Optoelectronic Matrix Switch Using Heterojunction Switching Photodiodes, Journal of Quantum Electronics, 1981.

Hayward, W., The Reliability of Telephone Traffic Load Measurements by Switch Counts, Bell System Technical Journal *31*, 357-377, 1952.

Hayward, W., Traffic Engineering and Administration of Line Concentrators, Proceedings of the 2nd International Teletraffic Congress, The Hague, 1958, paper 23.

Hayward, W., and Wilkinson, R., Human Factors in Telephone Systems and Their Influence on Traffic Theory Especially with Regard to Future Facilities, Proceedings of the 6th International Teletraffic Congress, Munich, 1970, pp. 431/1-10.

Herzog, U., A General Variance Theory Applied to Link Systems with Alternate Routing, 5th International Teletraffic Congress, New York, 1967, pp. 398-406.

Hieber, L., About Multi-Stage Link Systems with Queuing, 6th International Teletraffic Congress, Munich, 1970, pp. 233/1-7.

Hobbs, M., *Modern Communications Switching Systems*, Tab Books, Blue Ridge Summit, Pennsylvania, 1974.

Holtzman, J., Analysis of Dependence Effects in Telephone Networks, Bell System Technical Journal *50*, 2647-2662, 1971.

IEEE, *Standard Definitions of Tones for Communication Switching*, IEEE Standard 312-1977, The Institute of Electrical and Electronics Engineers, Inc., New York, 1977.

ISO Standards for Flow Charts, ISO/R, Geneva, Switzerland, 1969, p. 1028.

Iverson, V., *Analysis of Real Teletraffic Processes Based on Computerized Measurements*, 1973.

Jacobaeus, C., A Study on Congestion in Link Systems, Ericsson Technical *48*, 1-70, 1950.

Joel, A., An Experimental Switching System Using New Electronic Techniques, The Bell System Technical Journal *37*, 1091, September 1958.

Joel, A., What is Telecommunication Circuit Switching?, Proceedings of the IEEE *65*, No. 9, pp. 1237-1253, September 1977.

Kappell, J., Non-Blocking and Nearly Non-Blocking Multi-Stage Switching Arrays, 5th International Teletraffic Congress, New York, 1967, pp. 238-241.

Kawasaki, B., and Hill, K., Low Loss Access Coupler for Multimode Optical Fiber Distribution Networks, Applied Optics *16*, No. 7, 1794-1795, 1977.

Kegl, H., and Neovius, G., Electronic Private Branch Exchange ASD 551, Ericsson Review 4, 109, 1974.

Kharkevich, D., An Approximate Method for Calculating the Number of Junctions in a Crossbar System Exchange, Elektrosvyaz 2, 55-63, 1959.

Khinchine, A., *Mathematical Methods in the Theory of Queuing*, Griffin, 1960.

Kleinrock, L., *Queuing Systems*, Vols. 1 and 2, John Wiley and Sons, New York, 1975.

Kosten, L., Application of Artificial Traffic Methods to Telephone Problems, Teleteknik (English edition) 1, 107-110, 1957.

Kosten, L., On the Validity of the Erlang and Engset Loss Formulae, PTT-Bedr 2, 42-45, 1948.

Kosten, L., Manning, J., and Garwood, F., On the Accuracy of Measurements of Probabilities of Loss in Telephone Systems, Journal of the Royal Statistical Society *B11*, 54-67, 1949.

Kruithof, J., *Telephone Traffic Calculus*, Bell Telephone Manufacturing Company, Antwerp, January 1952.

Kruithof, J., Loss Formulas for Homogeneous Gradings of the Second Order in Telephone Switching Employing Random Hunting, Electrical Communications 35, 57-68, 1958.

Kuhn, P., Waiting Time Distributions in Multi-Queue Delay Systems with Gradings, Proceedings of the 7th International Teletraffic Congress, Stockholm, 1973, pp. 242/1-9.

Kummerle, K., An Analysis of Loss Approximations for Link Systems, 5th International Teletraffic Congress, New York, 1967, pp. 327-336.

Kyme, R., A System for Telephone Traffic Measurement and Routing Analysis by Computer, Proceedings of the 7th International Teletraffic Congress, Stockholm, 1973, pp. 525/1-6.

Lederman, S., and Petschenik, N., Automated Handling of International Calls Through TSPS No. 1, Conference Record IEEE International Conference on Communications, 1974, p. 11.

Leigh, R., Standards for the Economic Provision of Direct Routes, Proceedings of the 7th International Teletraffic Congress, Stockholm, 1973, pp. 522/1-5.

Local Telephone Networks, The International Telecommunications Union, Geneva, 1968.

Longley, H., The Efficiency of Gradings, Post Office Electrical Engineering Journal *41*, 45-49, 67-72, 1948.

Lotze, A., Traffic Variance Method for Gradings of Arbitrary Type, Proceedings of the 4th International Teletraffic Congress, London, 1964, paper 80.

Lotze, A., History and Development of Grading Theory, 5th International Teletraffic Congress, New York, 1967, pp. 148-161.

Lotze, A., Optimum Link Systems, 5th International Teletraffic Congress, New York, 1967, pp. 242-261.

MacDonald, R., and Hara, E., Optoelectronic Broadband Switching Array, Electronic Letters 14, No. 16, 502-503, August 1978.

MacDonald, R., and Hara, E., Switching with Photodiodes, IEEE Journal of Quantum Electronics QE16, No. 3, 289-295, 1980.

Marrows, B., Circuit Provision for Small Quantities of Traffic, Telecommunications Journal of Australia 11, 208-211, 1959.

Martin, N., A Note on the Theory of Probability Applied to Telephone Traffic Problems, Post Office Electrical Engineering Journal 16, 237-241, 1923.

McHenry, C., An Investigation of the Loss Involved in Trunking from Primary Line Switches to 1st Selectors via Secondary Line Switches in Strowger Automatic Exchanges, Post Office Electrical Engineering Journal 14, 217-227, 1922.

Method of Calculating Trunking and Switch Quantities for Strowger Automatic Telephone Exchanges, Bulletin 485, Automatic Electric Company, Chicago, Illinois, 1953.

Mina, R., The Theory and Reality of Teletraffic Engineering, Telephony, April 1971.

Molina, E., The Theory of Probabilities Applied to Telephone Trunking Problems, Bell System Technical Journal 1, 69-81, 1922.

Molina, E., Poisson's Exponential Binomial Limit, Van Nostrand, New York, 1947.

Murray, L., One-at-a-Time Operation in Telephone Exchanges, ATE Technical Journal 14, 256, 1958.

National Telephone Networks for the Automatic Service, International Telecommunications Union-CCITT, Geneva, 1964.

Networks, Laboratorios ITT de Standard Electrica SA, Madrid, 1973 (limited circulation).

Notes on Distance Dialling, American Telephone and Telegraph Company, New York, 1975.

Notes on Transmission Engineering, 1st edition, United States Independent Telephone Association, St. Louis, Missouri, 1963.

Numbering, Telecommunication Planning, ITT Laboratories, Madrid, 1973.

Nunoshita, M., Nomusa, Y., Matsui, T., and Nakayama, T., Optical Switch for Multimode Optical Fiber Systems, Optical Letters 4, 27-28, 1979.

Nyguist, H., Certain Factors Affecting Telegraph Speed and
    Certain Topics in Telegraph Transmission Theory, Transac-
    tions of American Institute of Electrical Engineers, 1924;
    1928.
O'Dell, G., An Outline of the Trunking Aspect of Automatic
    Telephony, Journal of the IEEE 65, 185-222, 1927.
Ogiwasa, H., and Suzuki, K., Optical Switching Experiment,
    IEEE Transactions in Communication COM. 27, 517-521, 1979.
Olsson, K., Anderberg, M., and Lind, G., Report on the 7th
    International Teletraffic Congress in Stockholm, Ericsson
    Technical 29, 107-144, June 1973.
Optimization of Telephone Trunking Networks with Alternate
    Routing, ITT Laboratories of Standard Electrica, Madrid,
    1974 (limited circulation).
Palm, C., Table of the Erlang Loss Formula, L. M. Ericsson,
    Stockholm, 1954.
Parviala, A., Calculation of the Optimum Number of Trunk Lines
    Based on Moe's Principle, Proceedings of the 7th International
    Teletraffic Congress, Stockholm, 1973, pp. 422/1-5.
Pearce, J., Logic Circuits, Telephony 49-55, October 1964.
Pearce, J., Logic Circuits (continued), Telephony 56-61, Novem-
    ber 1964.
Pearce, J., Electronic Switching, Telephony Publishing Company,
    Chicago, 1968.
Pearce, J., The New Possibilities of Telephone Switching, Pro-
    ceedings of the IEEE 65, September 1977.
Pearce, J. Gordon, Telecommunications Switching, Plenum Press,
    New York, 1981.
Povey, J., A Study of Traffic Variations and a Comparison of
    Post-Selected and Time-Consistent Measurements of Traffic,
    5th International Teletraffic Congress, New York, 1967,
    pp. 1-6.
Povey, J., and Cole, A., The Use of Electronic Digital Computers
    for Telephone Traffic Engineering, Post Office Electrical En-
    gineering Journal 58, 203-209, 1965.
Pratt, C., A Group of Servers Dealing with Queuing and Non-
    Queuing Customers, Proceedings of the 6th International Tele-
    traffic Congress, Munich, 1970, pp. 335/1-8.
Rapp, Y., The Economic Optimum in Urban Telephone Network
    Problems, Ericsson Techniques 49, 1-132, 1950.
Rapp, Y., Some Economic Aspects on the Long-Term Planning of
    Telephone Networks, Ericsson Review 45, 61-71, 122-136, 1968.
Reference Data for Radio Engineers, Chapter 36, Switching Net-
    works and Traffic Concepts, 6th edition, Howard W. Sams &
    Company, Inc., Indianapolis, 1975.

Reference Tables Based on A. K. Erlang's Interconnection Formula, Siemens & Halske Aktiengesellschaft, Munich, 1961.

Report of the 1st International Congress on the Application of the Theory of Probability in Telephone Engineering and Administration, Copenhagen, 1955, Teleteknik (English edition) *1*, 1-130, 1957.

Report of the 2nd International Congress on the Application of the Theory of Probability in Telephone Engineering and Administration, The Hague, PTT-Bedr *9*, 159-209, 1958.

Report on the 3rd International Teletraffic Congress, Paris, 1961, Annals of Telecommunication *17*, 145-226, 1962.

Report on the Proceedings of the 4th International Teletraffic Congress, London, Post Office Telecommunications Journal special issue, 1-66, 1964.

Riordan, J., *Stochastic Service Systems*, John Wiley and Sons, New York, 1962.

Robinson, N., Trunk Line Switching with Special Reference to the Separate Signaling System, ATM Technical Society paper 206, 1953.

Rodriguez, A., and Dartois, J., Traffic Unbalances in Small Groups of Subscribers, Electrical Communications *43*, 173-180, 1968.

Ronnblom, W., Traffic Loss of a Circuit Group Consisting of Bothway Circuits Which Is Accessible for the Internal and External Traffic of a Subscriber Group, Teleteknik 79-92, 1959.

Rubas, J., Analysis of Congestion in Small PABX's, Proceedings of the 6th International Teletraffic Congress, Munich, 1970, pp. 211/1-8.

Rubin, M., and Haller, C., *Communication Switching Systems*, Reinhold, 1966.

Ryan, J., The Role of the ITU and CCITT in Telecommunications, International Switching Symposium, 1974, pp. 121/1-121/7.

Saaty, T., *Elements of Queuing Theory with Applications*, McGraw-Hill, New York, 1961.

Sinkhorn, R., Diagonal Equivalence to Matrices with Prescribed Row and Column Sums, American Mathematical Monthly *74*, 401-405, 1967.

Smith, N., More Accurate Calculation of Overflow Traffic from Link-Trunked Crossbar Group Selectors, Proceedings of the 3rd International Teletraffic Congress, 1961, Paris, paper 36.

Smith, N., Erlang Loss Tables and Other Parameters for Normally Distributed Offered Traffics, Proceedings of the 4th International Teletraffic Congress, 1964, paper 100.

Smyth, D., An Adaptive Forecasting Technique and Traffic Flow Matrix Forecasting, Australian Telecommunications Research *3*, 38-42, 1969.

Spenser, A., and Vigilante, F., No. 2 ESS System Organization
    and Objectives, The Bell System Technical Journal *48*, 2615,
    October 1969.

Switching Systems, American Telephone and Telegraph Company,
    New York, 1961.

Syski, R., *Introduction to Congestion Theory in Telephone Sys-
    tems*, Oliver and Boyd, London, 1960.

Thierer, M., Delay Systems with Limited Accessibility, 55th In-
    ternational Teletraffic Congress, New York, 1967, pp. 203-213.

Tomlin, J., and Tomlin, S., Traffic Distribution and Entropy,
    Nature *220*, 974-976, 1968.

*Traffic Engineering Practices — Trunk Facilities — Basic Trunk
    Tables*, American Telephone and Telegraph Co., Div. G,
    Section 5-a, March 1960.

Van Den Bossche, M., and Knight, R., *Traffic Problems and
    Blocking in a Three-Link Switching System*, Bell Telephone
    Manufacturing Company, Antwerp, February 1957.

Vaughan, H., Introduction to No. 4 ESS, International Switch-
    ing Symposium, 1972, pp. 19-25.

Wagner, W., On Combined Delay and Loss Systems with Nonpre-
    emptive Priority Service, 5th International Teletraffic Congress,
    New York, 1967, pp. 73-84.

Wallstrom, B., Artificial Traffic Trials in a Two-Stage Link Sys-
    tem Using a Digital Computer, Ericsson Technical *14*, 259-289,
    1958.

Warman, J., and Bear, D., Trunking and Traffic Aspects of a
    Sectionalised Telephone Exchange System, Proceedings of the
    IEEE *113*, 1331-1343, 1966.

Wikell, G., On the Consideration of Waiting Times of Calls De-
    parted from the Queue of a Queuing System, 5th International
    Teletraffic Congress, New York, 1967, p. 498 (abstract).

Wikell, G., Manual Service Criteria, Proceedings of the 6th In-
    ternational Teletraffic Congress, Munich, 1970, pp. 134/1-6.

Wilkinson, R., The Interconnection of Telephone Systems —
    Graded Multiples, Bell System Technical Journal *10*, 531-564,
    1931.

Wilkinson, R., Discussion Contribution on Basic Theory Under-
    lying Bell System Facilities Capacity Tables, by A. L. Gracey,
    AIEE Transactions 238-244, 1950.

Wilkinson, R., *Working Curves for Delayed Exponential Calls
    Served in Random Order*, Bell System Technical Journal *32*,
    360-383, 1953.

Wilkinson, R., *Theories for Toll Traffic Engineering in the USA*,
    Bell System Technical Journal *35*, 421-514, 1956.

Wilkinson, R., *Non Random Traffic Curves and Tables for Engineering and Administrative Purposes*, Traffic Studies Center, Bell Telephone Laboratories, 1970.

Wilkinson, R., Some Comparisons of Load and Loss Data with Current Teletraffic Theory, Bell System Technical Journal *50*, 2808-2834, 1971.

Young, W., and Curtis, L., Moving Optical Fiber Array Switch for Multimode and Single-Mode Fibers, 3rd IOOC Conference, April 27-28, 1981.

II

OPTOELECTRONIC PROCESSORS

5

Hardware of Optoelectronic Processors

5.1 BASIC LOGIC OPERATIONS IN COMPUTERS

As may be seen from the block diagram in Fig. 5.1, digital computers comprise the following basic units:

Arithmetic unit
Memory unit
Control unit
Data input unit
Result output unit

which are functionally independent. Let us discuss their functions.

The *arithmetic unit* executes numerical data arithmetic, logic, and other operations into which any computation may be decomposed. It is built around an adder that sums two numbers. All other operations may be reduced to single or multiple addition and some auxiliary actions (e.g., shifts required for multiplication). The arithmetic unit is characterized by speed, i.e., the time required for the execution of an operation, or by the number of operations per unit time, as well as by the set of executable operations.

Computer memory stores all necessary information: source data and partial and final results. It also stores task execution algorithms. The memory receives data sent by other units (e.g., by the arithmetic unit) or coming to the computer through input units, and supplies all other units with data required by computations.

Computer memory is characterized by size (capacity) and speed (or access time). Memory speed is defined by the access time, i.e., by the time required to read in or out of the memory.

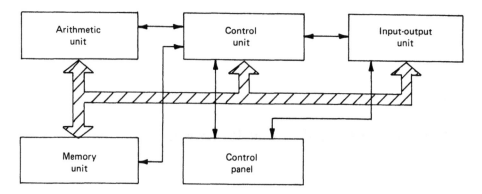

Figure 5.1. Block diagram of electronic digital computer.

Naturally, it is also desirable to have as large a memory as possible. However, one may easily see that these two requirements
are inconsistent: the larger the memory, the longer the read/
write time. This contradiction is eliminated by decomposition of
the memory into main (internal) and mass (external) memories.

The main memory in each computer is decomposed into separate cells consisting of some fixed number of positions. Each
position may take one of two possible values 0 or 1. Thus, memory cell content is regarded as a binary number, which is termed
a *computer word*. Memory capacity is characterized by the number of computer words that may be stored in it.

The main memory directly communicates with other computer
units, including those of control and arithmetic. The *input* unit
transforms source data into a convenient form and sends all the
necessary data into the memory. This information comes directly
only to the main memory. The *output* unit transforms computation results into a form suitable to human perception and further
use. The *control unit* provides interaction of all computer units,
thus automatically controlling all the computations. The characteristics of current super-high-speed computer systems are listed
in Table 5.1. The solution of a problem lies in the execution of
an *algorithm*, i.e., a certain prescribed succession of operations.
This notion involves not only arithmetic or algebraic operations,
but also other types of actions, namely, comparison of numbers,
transfer to the next operation, conditional actions that are executed or omitted depending on some conditions, etc.

Table 5.1 Characteristics of High-Speed Computer Systems

Manufacturer	Cray Research	Control Data	Texas Instruments
System	Cray-1	STAR-100	ASC
Length of computer word, bits	64	64	64
Maximal operation execution speed, M 64-bit words/sec			
Addition	80	50	37
Multiplication	80	25	21
Division	80	12,5	18
Capacity of primary memory, M words	0,262-1	0,524-1	0,131-1
Cycle time, n sec	50	1280	160

In order to enable algorithm execution by a digital computer, one has to represent it as a succession of elementary computer operations. The algorithm of a given problem represented as a succession of elementary computer operations is referred to as the *program*.

Today's digital computers make use of binary notation, which requires only two figures. Therefore, numbers in this system may be represented by bistable elements, one of the states representing 0, the other one representing 1. Memory elements must store one binary digit, and the arithmetic unit performs arithmetic operations in binary notation, which makes them simple and obvious. Generally, each computer word, independent of its function, is a set of binary digits, i.e., of 0s and 1s.

Any computer operation results in a binary number; therefore, logical elements of the computer may be regarded as functional transformers with binary numbers at their inputs and outputs. Separate bits may be conveniently regarded as arguments and values for the function. In this case the logic element takes

Arguments			Function
a	b	c	z
0	0	0	1
0	0	1	0
0	1	0	1
0	1	1	0
1	0	0	0
1	0	1	0
1	1	0	1
1	1	1	1

Figure 5.2. Tabular definition of a function of three variables:
an example.

the form of a multiinput and multioutput device, the binary digits
0 or 1 fed into each input and generated at each output.

Thus, we deal with variables and functions taking the values
0 and 1 only. They are called *logic* or *Boolean* after the English
mathematician and logician George Boole (1815-1864), the founder
of the modern symbolic logic.

Arguments of logic functions may take only two different values. Therefore, there are a finite number of different argument
sets for a function of any number of variables, and logic function
may be defined by a table listing all the possible argument sets
and corresponding function values. The function of three variables $z = f(a, b, c)$ may be defined by an eight-row table (Fig.
5.2); the function of n variables has 2^n different argument sets.
Tables are cumbersome and do not adequately visualize multiple
arguments. It is more convenient to express logic functions by
means of formulas in terms of simpler and better known functions
of smaller numbers of arguments. Consider the basic logic functions of two arguments.

Logic AND

The values for AND of two variables are defined as in Fig. 5.3,
which shows that they coincide with the argument product. That

Arguments		Function
a	b	AND
0	0	0
0	1	0
1	0	0
1	1	1

Figure 5.3. Truth table of logic function AND.

is why this operation is called *logic multiplication*; in mathematical logic this operation is termed conjunction. It is easy to define AND of an arbitrary finite number of arguments; it is 1 if and only if all the factors are 1.

Logic product is denoted by Λ. It obeys the laws of commutativity and associativity:

$a \Lambda b = b \Lambda a$ commutativity

$(a \Lambda b) \Lambda c = a \Lambda (b \Lambda c)$ associativity

Logic OR

This function of two arguments is defined by the table in Fig. 5.4. It is referred to as *logic addition* or *disjunction* and is denoted by V. It might be well to point out that the logic addition of two bits

Arguments		Function
a	b	OR
0	0	0
0	1	1
1	0	1
1	1	1

Figure 5.4. Truth table of logic function OR.

Argument	Function NOT
0	1
1	0

Figure 5.5. Truth table of logic function NOT.

differs from their arithmetic sum. This operation also obeys the commutative and associative laws.

$a \lor b = b \lor a$

$(a \lor b) \lor c = a \lor (b \lor c)$

Moreover, the distributive law holds for logic addition and multiplication:

$a \land (b \lor c) = (a \land b) \lor (a \land c)$

The logic sum of several terms is zero if and only if all of them are zeros.

Logic Negation (NOT)

Logic negation depends on one argument and is defined as in Fig. 5.5. It is denoted by a barred argument, $\bar{a}$. Like arithmetic operations, logic operations have different "priorities," which are reflected by the order of their execution in expressions: first, NOT is executed, followed by AND, and only after them is OR performed. This is completely similar to algebraic expressions. When computing the formula $a^2 b^2 + c^2$, one first raises to the second power, then multiplication $a^2 b^2$ and only then addition. Similarly, in the formula $\bar{a} \land \bar{b} \lor \bar{c}$, first negations are performed, next conjunctions, and after them disjunctions. In cases where this order should be changed, parentheses are used according to the same rules as in algebraic formulas.

Logic functions enable one to substitute simple formulas for cumbersome tables. For example, the function of three arguments $z = f(a, b, c)$ given in Fig. 5.2 may be written as the simple formula

$z = \bar{a} \land \bar{c} \lor a \land b$

Similar representation is possible for other logic functions. Let the logic function p of three arguments $u = p(a, b, c)$ be defined by those argument sets where it is 1:

$a = 1$ $b = 0$ $c = 1$

$a = 1$ $b = 1$ $c = 0$

$a = 0$ $b = 0$ $c = 1$

Assign to each set a logic product of arguments where each zero cofactor has the sign of negation, and obtain three functions:

$F_1(a, b, c) = a \wedge \bar{b} \wedge c$

$F_2(a, b, c) = a \wedge b \wedge \bar{c}$

$F_3(a, b, c) = \bar{a} \wedge \bar{b} \wedge c$

The logic sum of the three products results in the following formula.

$p = F_1 \vee F_2 \vee F_3 = a \wedge \bar{b} \wedge c \vee a \wedge b \wedge \bar{c} \vee \bar{a} \wedge \bar{b} \wedge c$

Direct checking readily demonstrates that this formula defines just the required function.

A similar construction results in a function defined by the sets of all those arguments where it is zero. Let, for instance, these sets be as follows:

$a = 0$ $b = 1$ $c = 1$

$a = 1$ $b = 0$ $c = 0$

$a = 0$ $b = 0$ $c = 0$

Construct the sums $a \vee \bar{b} \vee \bar{c}$, $\bar{a} \vee b \vee c$, and $a \vee b \vee c$, and form their logic product

$F = (a \vee \bar{b} \vee c) \wedge (\bar{a} \vee b \vee c) \wedge (a \vee b \vee c)$

One can easily see that this function is zero over given argument sets and only over those sets.

Thus, we ensured that any logic function may be represented as a combination of ANDs, ORs, and NOTs. Such a system of logic functions, sufficient for construction of any other function, is referred to as *functionally complete*. It is not unique; other functionally complete systems are also possible.

The *Sheffer stroke* is one of the functionally complete logic functions. It is defined by the table in Fig. 5.6 and is represented as $F(a, b) = \overline{(a \wedge b)} = \bar{a} \vee \bar{b}$. Another example of functional completeness is represented by the *Peirce function*,

Arguments		Function
a	b	F (a, b)
0	0	1
0	1	1
1	0	1
1	1	0

Figure 5.6. Truth table of logic function Sheffer stroke.

defined by the table in Fig. 5.7 and represented as $F(a, b) = \overline{(a \lor b)} = \bar{a} \land \bar{b}$. Each makes a functionally complete system, i.e., any logic function may be expressed through either of them. The most frequently used logic functions are shown in Fig. 5.8.

A comparison of the functions F_2 through F_9 shown in Fig. 5.8 demonstrates that F_2 and F_6, F_3 and F_7, F_4 and F_8, and F_5 and F_9 are mutually inverse; i.e., they are negations of each other and take opposite values over the same sets of variables:

$$F_2 = \overline{F}_6 \quad \overline{F}_2 = F_6 \quad \text{or} \quad x_1 \lor x_2 = \overline{x_1 \downarrow x_2} \quad \overline{x_1 \lor x_2} = x_1 \downarrow x_2$$

$$F_3 = \overline{F}_7 \quad \overline{F}_3 = F_7 \quad \text{or} \quad x_1 x_2 = \overline{x_1 \uparrow x_2} \quad \overline{x_1 x_2} = x_1 \uparrow x_2$$

$$F_4 = \overline{F}_8 \quad \overline{F}_4 = F_8 \quad \text{or} \quad x_1 \rightarrow x_2 = \overline{x_1 \leftarrow x_2} \quad \overline{x_1 \rightarrow x_2} = x_1 \leftarrow x_2$$

$$F_5 = \overline{F}_9 \quad \overline{F}_5 = F_9 \quad \text{or} \quad x_1 \infty x_2 = \overline{x_1 \oplus x_2} \quad \overline{x_1 \infty x_2} = x_1 \oplus x_2$$

The above relations demonstrate the feasibility of expressing logic functions through other logic functions. Functions F_1 through F_9

Arguments		Function
a	b	F (a, b)
0	0	1
0	1	0
1	0	0
1	1	0

Figure 5.7. Truth table of logic Peirce function.

Functions	Arguments					
	$x = 0$	$x = 1$	$x_1 = 0$ $x_2 = 0$	$x_1 = 1$ $x_2 = 0$	$x_1 = 0$ $x_2 = 1$	$x_1 = 1$ $x_2 = 1$
Negation $F_1(x) = x$	1	0	–	–	–	–
Disjunction $F_2(x_1, x_2) = x_1 \lor x_2$	–	–	0	1	1	1
Conjunction $F_3(x_1, x_2) = x_1 x_2$	–	–	0	0	0	1
Implication $F_4(x_1, x_2) = x_1 \rightarrow x_2$	–	–	1	0	1	1
Equivalence $F_5(x_1, x_2) = x_1 \sim x_2$	–	–	1	0	0	1
Peirce function $F_6(x_1, x_2) = x_1 \downarrow x_2$	–	–	1	0	0	0
Sheffer stroke $F_7(x_1, x_2) = x_1 \uparrow x_2$	–	–	1	1	1	0
Inhibition $F_8(x_1, x_2) = x_1 \leftarrow x_2$	–	–	0	1	0	0
Inequivalence $F_9(x_1, x_2) = x_1 \oplus x_2$	–	–	0	1	1	0

Figure 5.8. Table of logic functions most popular in computer engineering.

are elementary and enable the representation of any composite logic function as an equivalent set of elementary functions.

Listed below are examples of functionally complete systems:

$F_2 = x_1 \lor x_2 \qquad F_3 = x_1 x_2 \qquad F_1 = \overline{\overline{x}}$

$F_2 = x_1 \lor x_2 \qquad F_1 = \overline{x}$

$F_3 = x_1 x_2 \qquad F_1 = \overline{\overline{x}}$

$F_6 = x_1 \downarrow x_2$

$F_7 = x_1 \uparrow x_2$

The first system is used in many cases as a functionally complete basis. A set of elements corresponding to any functionally complete system of elementary logical functions is referred to as a functionally complete set of logical elements.

The names of logical elements reflect their functions. Negation is implemented by the inverter (NOT); conjunction is implemented by AND; disjunction requires OR; the Sheffer stroke is implemented by the Sheffer element (NAND); and the Peirce operation is implemented by the Peirce element (NOR).

In electronic computers, binary operations are represented by electrical signals, but optoelectronic processors use light for data transmission. The following binary representation is most commonly used:

Light: 1
No light: 0

For the sake of illustration, consider the execution of basic logic operations by an electrically controlled space-time light modulator (SLM). It is a discrete light gate in the form of matrix, each of whose elements executes logic operation over one data bit. The transmission of each element is controlled by an electrical field. With no field applied, all the cells have identical optical properties. But if electrical potentials are applied to optically transparent electrodes deposited on flat SLM surfaces, the optical characteristics of the SLM material change. These are denoted by T and $\bar{T}$ SLM transmitting light under signal ($x = 1$) and without signal, respectively. Figure 5.9 demonstrates methods of executing various logical operations by means of space-time light modulators. The schemes shown in Fig. 5.10 perform multiplace conjunction $F = \wedge_{i=1}^{n} x_i$, multiplace disjunction $F = \vee_{i=1}^{m} x_i$, and the Boolean function

$$F = \bigvee_{i=1}^{m} \left(\bigwedge_{j=1}^{n} x_{ij} \right)$$

5.2 MAJOR REQUIREMENTS FOR LOGIC GATES OF OPTOELEC-TRONIC DIGITAL COMPUTER SYSTEMS

The design philosophies of digital optoelectronic computer systems may differ significantly and depend essentially on the particular set of elements performing discrete computational operations over optical signals. The major problem here lies in finding the optimal trade-off among physical principles, materials and technology for

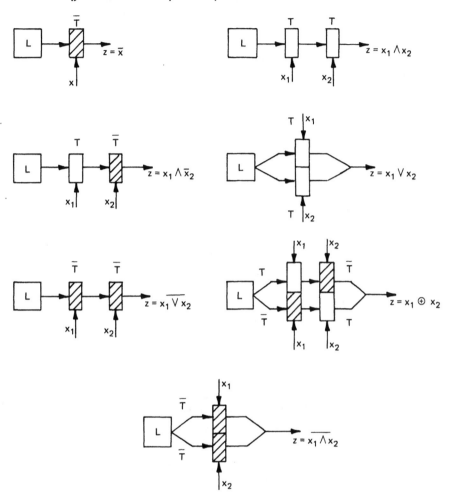

Figure 5.9. Examples of logic operations implemented by electrically controlled SLM. Modulators passing light in the presence (x = 1) or in the absence of control signal are denoted by T and $\bar{T}$, respectively.

implementation of elements required by these principles, and the materials and optoelectronic computer engineering. The state of the art is likely to satisfy the requirements for optoelectronic hardware and enable its development.

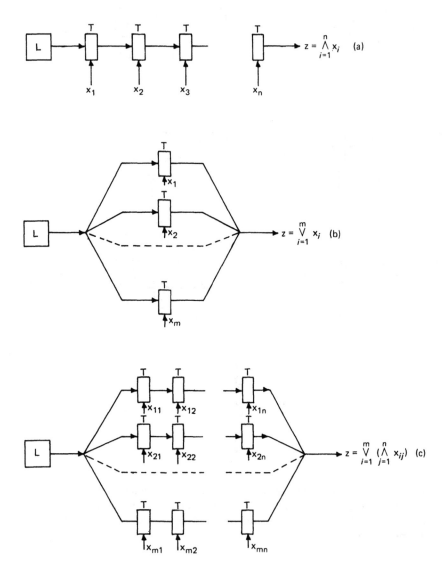

Figure 5.10. Examples of logic operations over multiple variables implemented by electrically controlled SLM: $z = \bigwedge_{i=1}^{n} x_i$; $z = \bigvee_{i=1}^{n} x_i$; and $z = \bigvee_{i=1}^{m} (\bigwedge_{j=1}^{n} x_{ij})$.

The experience gained by computer engineering enables formulation and generalization of the requirements of this new kind of hardware. All the requirements may be classified into four groups:

1. Requirements for the logic elements
2. Physical requirements for the elements
3. System engineering requirements
4. Technological requirements

The requirements for logic elements that provide a complete set of logic operations, data storage, and reproduction stipulate the availability of the following elements:

1. Signal paths allowing signal fan-in and fan-out
2. Threshold elements discriminating quantized signal levels (0 and 1 for the binary code)
3. Logic elements implementing a complete system of logic functions, i.e., at least one of the following four sets of logic elements: conjunction (AND) and negation (NOT); disjunction (OR) and negation (NOT); Sheffer stroke (NAND); and Peirce function (NOR)
4. Memory elements enabling data writing, storage, and reproduction
5. Amplifiers restoring signal power up to the standard level

The negation element and amplifier are necessary to a general-purpose data processing system of elements. Without the NOT element, only a narrow class of analog signal processing is feasible, and general-purpose logic processing becomes impossible. Without the amplifier, only a narrow class of systems without feedback and with a limited number of processing stages may be implemented.

The physical requirements may be satisfied by various optical and photoelectric phenomena, characterized as follows:

1. An effective signal gain providing great values of the gain band B, i.e., the product of gain G by bandwidth Δf
2. A strong and quickly responding nonlinearity of signal transformation enabling a sharp and quick response threshold of the logic element
3. Physical phenomena that may be used for development of a memory element highly sensitive to the control signal and quickly responding to it
4. A large bandwidth of data paths between elements
5. Low power consumption per bit of processed information, i.e., the product of control signal power by response time of logic and memory elements, i.e., by high power quality expressed as the number of operations per unit power

6. The possibility of satisfying requirements 1, 2, and 3 in small physical volumes

System engineering requirements enable unification of separate elements into circuits performing the desired signal processing functions. The following are necessary to this end.

1. Identical circuit elements
2. Standard signal representation and identical (with certain tolerances) input and output parameters of logic elements
3. A set of engineering means for the unification of separate elements into functional units
4. The exclusion of element interdependence in a circuit, i.e., their effective decoupling ensuring stability of elements to cross fire
5. The reliability and stability of the elements in circuits, as well as sufficient lifetime

The technological requirements boil down to the possibility of meeting the engineering requirements:

1. Materials with parameters required for elements and circuits
2. Technological processes yielding sufficiently good materials
3. Technologies for the unification of elements into circuits
4. Element reproducibility (including parameter scatter) assuring an economically acceptable output of finished product
5. Technological equipment enabling one to satisfy conditions 1 through 4

Many of the above requirements are well known and might seem trivial, but it is very difficult to meet them in toto. Violation of any of them might make new hardware unsuitable or at least uncompetitive, no matter how attractive seem the principles underlying it.

In view of the fact that effective purely optical active elements for light amplification and nonlinear transformation are still not available, high-throughput optical data processing devices and systems may be designed mostly by finding a good trade-off between passive optical elements transforming coherent light signals with active semiconductor electronic devices. This approach allows one to estimate the limiting speed of optical digital data processing.

The power quality of amplifiers and analog elements is defined as

$$K = G^{1/2} \frac{\Delta f}{P}$$

where P is the power dissipated by the element, G is the power gain, and Δf is bandwidth. The power quality K is the inverse of power per logical operation; i.e., it is the number of logical operations of an element per unit power.

For elements of modern integrated circuits, values $K = 10^{11}$ to 10^{13} J^{-1} have been attained, the limiting value of quality being $K_0 = 10^{16}$ J^{-1}. The importance of K is due to the fact that it defines the maximal allowable density of data flow under limited dissipated power. The temperature limit is defined by the maximal heat Q dissipation from the surface of, say, an integrated circuit, which under natural cooling is characterized by $Q \sim 1$ W/cm^2. Under a given element switching frequency f and gain G, the power quality K defines the upper bound of element density over the circuit plane, i.e.,

$$N \leq \frac{KQ}{fG} S$$

If N elements comprise an optoelectronic circuit occupying area S, its throughput is defined by the following simple relation:

$$I_0 = Nf = \frac{SQK}{G} \tag{5.1}$$

whence, assuming that $K = 10^{13}$ J^{-1}, $Q \sim 1$ W/cm^2, and $S = 10$ cm^2 are practically attainable and that gain $G \sim 10$ to 10^2 is sufficient for modulation of optical properties of passive optical elements, one gets a maximal throughput of 10^{12} to 10^{13} bps (bits per second).

The finiteness of speeds of light and electrical signals in control circuits imposes some constraints because of the necessity to synchronize the operation of elements of optoelectronic digital computer systems. For linear dimensions of lenses and space-time light modulators equal to 10 cm, path differences in the optical system are of the order of optical element dimensions, i.e., about 10 cm. Assuming that the time mismatch for various system elements should not exceed some fraction γ of the cycle period $T = f_T^{-1}$ obtain for $\gamma = 1/30$ that

$$f_T = \frac{\gamma C}{L} = 100 \text{ Mcps}$$

Hence, in an optoelectronic digital computer system with marginal parameters, a typical data processing rate should be about 10^8 cps and maximal number format $N_{max} = 10^5$ to 10^6.

Interestingly, for a practical number format $N = 10^4$ and space-time light modulator area $S = 10$ cm^2, the size of one image cell is about 300 μm, which is far from the diffraction limit $\lambda / \sqrt{\Omega}$, where λ is the light wavelength and Ω is the lens angular aperture. For such a picture, transmission redundancy is as follows:

$$R = \frac{S\Omega}{\lambda^2 m^2 N}$$

where m is the greatest diffraction order required for good transmission of the binary image of a number. For $\lambda = 6.10^{-5}$ cm, $\Omega = 10^{-1}$ sr, $N = 10^4$, $S = 10$ cm^2, and $m = 3$, a lens 4 cm in diameter provides redundancy $R = 3 \times 10^3$, thus assuring great margins for aberration tolerances and high noise immunity of information transmission.

It should be noted that data I/O (input/output) is of prime importance in computer circuits. If signals enter and leave the circuit through its periphery, as is the case with electronic integrated circuits, the throughput of circuits with a sufficient number of elements N is proportional to the I/O rate, i.e., $I_e \sim f_T N^{1/2}$ because the number of output channels is proportional to $\sqrt{N}$. Hence, for electronic integrated circuits,

$$I_e \sim b \ \frac{K_e Q S_e f_T}{G_e}^{1/2} \qquad\qquad (5.2)$$

where b is a geometrical factor.

In optoelectronic circuits, I/O channels are two dimensional, thus providing simultaneous access to all circuit elements. The circuit throughput I_0 is determined through equation (5.1). Comparison of the right sides of (5.1) and (5.2) reveals the principal difference between the limitations of electronic and optical computations. The frequency-independent right side of (5.1) indicates that a relative reduction in the speeds of optoelectronic computations may be offset, within reasonable boundaries, by the greater number of computation paths and elements in the circuit N.

The versions of basic digital logic elements shown in Fig. 5.11 rely upon semiconductor electronic elements and controlled passive optical elements. Figure 5.11b shows a two-input logical element performing functions AND and OR. The information about the logical variables x_1 and x_2 arrives at the photodiodes connected to an electronic logical gate performing one of these operations and amplifying the signal. The amplified signal controls the optical reflecting modulator, based, for example, on the linear electrooptic effect. External linearly polarized light falls on the modulator, which reflects it only if both input signals coincide (AND) or if there is at least one input signal (OR).

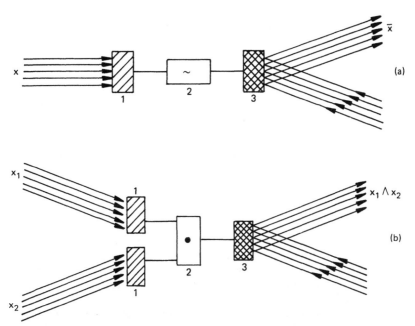

Figure 5.11. (a) Negation element. Electronically inverted and amplified signal from photodiode controls optical modulator: (1) photodiode, (2) inverter amplifying electric signal, and (3) light modulator. (b) Two-input logic element executing functions AND and OR. Signals from photodiodes are input into an electronic logic gate executing either operation. The amplified signal controls the optical reflecting modulator: (1) photodiodes, (2) logic AND, and (3) optical reflecting modulator.

Figure 5.11a schematically depicts the negation element. The input signal arrives at a photodiode transforming light signals into electrical signals that are electronically inverted. The output signal controls the reflection of the optical modulator. If no signal enters the circuit, the space-time modulator cell reflects the external optical signal; otherwise, the reflection coefficient is minimal (zero in the limit).

The threshold properties of photodiodes enable discrimination of the quantized signal levels, i.e., 0 and 1 in the binary code. The circuit amplifier restores signal power to the standard level, assuring control of the passive optical element. The power of

	1	2	3	4	5	6
A	+	+	+	$+/-$	+	
B	+	+	$+/-$	+	$+/-$	+
C	+	+	−	+	−	
D	$+/-$	+	−	−	$+/-$	

Figure 5.12. Requirements for optoelectronic digital processor hardware: state of the art.

the reflected signal is supplied by the external source of coherent light. The electronic circuitry executes a complete system of Boolean functions; it also involves a dynamic flip-flop, thus permitting the space-time light modulator to operate as a dynamic memory cell. In this case, the signal may be reflected even if there is no input signal.

Let us see to what extent digital optoelectronic logic elements meet the set of requirements for new hardware. In Fig. 5.12, the requirements that can be met are marked by +; those that are not thoroughly studied or were not considered at all are marked by -.

The issues involved in the development of electronic components, as well as numerous electrooptic and magnetooptic solid-state phenomena, have been well studied separately. As one may see from the figure, the design of high-speed matrix space-time light modulators integrating electronic and optical parts by means of integrated circuit (IC) technology is still to be considered.

It should be noted that, sometimes, a light gate may be replaced by a light-emitting element. A complicated matrix optron may be built around matrix photodiodes and other light-sensitive devices manufactured by means of microelectronic IC technology. One may anticipate a system with optron matrices of photodiodes, logic gates, and light-emitting diodes with about 10^4 elements and a cycle time about 10^{-7} sec, thus supporting the throughput of 10^{11} bps. The essential divergence of light-emitting diodes requires in this case the application of fiber optics for picture transmission and transformation.

5.3 HARDWARE FOR OPTOELECTRONIC DIGITAL PROCESSORS

Figure 5.13 shows a general block diagram of a multichannel digital optical data processing path. Input data to be processed I_1 are

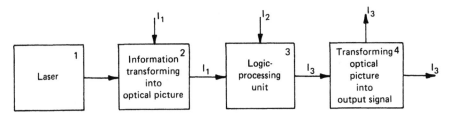

Figure 5.13. General scheme of a multichannel digital optical data processing path.

transformed in unit 2 into an optical picture. Another data flow I_2, defining the program processing I_1, enters the logic processing unit 3. The resulting data flow I_3 is transformed into a desired form by transmitting unit 4, which sends data for further processing and also transforms optical signals into electrical signals.

The data processing rate depends essentially on the speeds with which data flows I_1 and I_2 enter the logic processing unit. For the generation of pictures and their input into the optical processing path, two basic techniques support a sufficiently high input rate.

The first one (Fig. 5.14) is intended for data arriving as a succession of electrical signals through several input channels simultaneously. An electrically controlled matrix space-time light modulator forms pictures in a rowwise fashion and stores the data until reading out.

The second technique is used for the input of task execution algorithms, instructions, and constants stored by read-only (main) optical memory. Figure 5.15 shows how data are read out of the optical memory. They are stored in separate cells of a planar matrix in the form of holograms and are read out by means of a laser addressing unit that directs the beam to a desired cell. The address unit is a multipositional laser beam deflector with random addressing. The two-dimensional picture readout of a hologram is sent to the space-time light modulator with optical input and modulates its transmission or reflection function. Figure 5.16 demonstrates and block diagram of a digital optoelectronic processor. Thus, the following devices are required for implementation of digital multichannel optoelectronic processors:

1. Lasers.
2. Multipositional random access light beam deflectors.

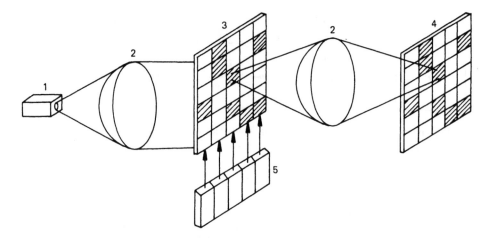

Figure 5.14. Data input into an optical data processing unit by means of electrically controlled space-time light modulator: (1) light source, (2) lenses, (3) multielement electrically controlled SLM, (4) input of the succeeding element to the data processing unit, and (5) SLM controller.

3. Electrically controlled multielement space-time light modulators with memory (ECSLM).
4. Optically controlled multielement space-time light modulators (OCSLM) executing AND, OR, and NOT, and amplifying the picture at the expanse of the light falling on the reflecting surface of the OCSLM. Of all operations, signal inversion and amplification are the most important, those of conjunction and disjunction implementable by circuit engineering methods.
5. Read-only and random-access optical accumulators, and memory media. In some cases, the optically controlled space-time light modulator with memory may operate as the optical main memory.
6. Optical elements supporting assembly, shift, masking, and transposition of digital pictures.

Conventional elements used for the generation of optical pictures (lenses, mirrors, polarizers, etc.) should also be included in this list. The above set of devices 1 to 6 forms a basis sufficient for designing general-purpose optical digital computers.

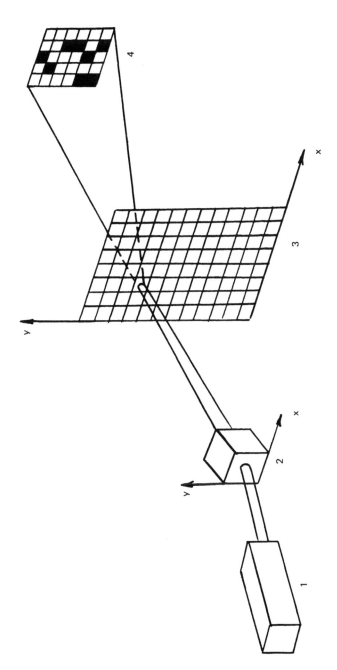

Figure 5.15. Data input into optical data processing unit by means of holographic memory: (1) laser, (2) multipositional deflector, (3) hologram matrix, and (4) input of data processing unit.

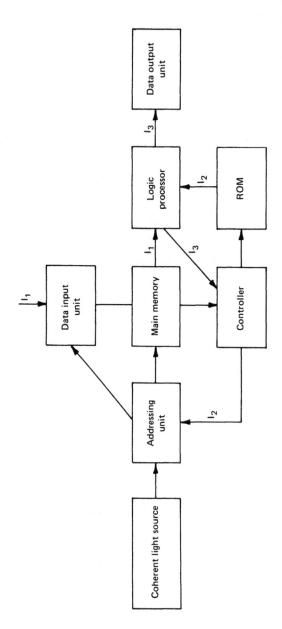

Figure 5.16. Block diagram of a digital optoelectronic processor.

Consider the state of the art of the basic units for optoelectronic processors.

Light Sources

Lasers generating sufficiently powerful, monochromatic and coherent light with low divergency are envisioned as light sources for optoelectronic processors. There is a very wide variety of lasers with very different physical and operational characteristics. Let us dwell upon some of them, bearing in mind the above requirements and the possibility of continuous generation.

Among the solid-state lasers, those based on YAG (yttrium aluminum garnet) that are activated by neodymium and generate light of wavelength 1.06 μm should be noted. High power is among their merits. However, they have such demerits as infrared operating wavelength, low efficiency (especially taking into account the transformation to the visual range), and poor coherence. Today, liquid lasers in which solutions of organic and nonorganic dyestuffs are usually used as active media are well developed. They feature the possibility of varying the frequency in a large range. But dye-based lasers are mostly insufficiently monochromatic and usually require powerful pumping from another laser, thus resulting in low total efficiency.

Therefore, solid-state and liquid lasers seem to be questionable for holographic processors. More promising are the gas and semiconductor lasers. Let us dwell upon the following types of gas lasers:

Lasers on atomic gases, as represented by helium-neon lasers
Ion lasers, exemplified by argon and krypton lasers
Lasers on metal vapors of which helium-cadmium, helium-selenium,
    and copper lasers are the most popular

The first of the gas lasers, helium-neon, is the best known and is widely used. It has generation lines in visual and infrared ranges, of which the red line at 0.63 μm is, naturally, the most interesting. The radiation coherence length is ~ 10 cm, the output power of continuous generation is ~ 10 mW with one transverse mode, and the laser efficiency is about several hundredths of 1%. Ion lasers make use of electron transitions of ions of argon, krypton, and other inert gases. Pumping is done by means of a high-current voltaic arc excited in the narrow laser pipe capillary. The most popular are the argon and krypton ion lasers, which feature high radiation power in continuous mode (1 to 100 W) and a sufficiently great coherence length for each individual line (~10 cm). Owing to the high pumping currents, pipe cooling is needed.

The helium-cadmium laser also has rather good performance, with a continuous generation power at wavelength 0.44 μm of tens and hundreds of milliwatts.

A special position among the lasers is occupied by semiconductor injection lasers, which differ both in physical processes and in possibilities of using them in optoelectronic processors. They effectively transform input power into coherent radiation. Injection lasers are very compact and compatible in power supply with semiconductor integrated circuits. Their generation raise time is about 10^{-9} sec. Injection lasers are sufficiently coherent for holographic memories. Injection laser rows and matrices enable the design of simpler radiation deflection systems. Currently, injection laser lifetime is over 10^4 h, and it is steadily growing. The mean radiation power in continuous single mode is 5 to 10 mW.

As follows from the comparison of basic characteristics of various laser types (Table 5.2), injection lasers are unique sources of coherent radiation for implementation of high-throughput digital optoelectronic systems, although experimental systems may be built with a rather wide variety of lasers.

Multipositional Deflectors

These are intended for changing laser beam spatial position according to a given law. In holographic memory systems, deflectors are used for random access to memory locations.

Deflectors may rely on electrooptic, acoustooptic, magneto-optic, and other physical phenomena.

Electrooptic discrete deflectors are multistep devices, each step consisting of polarizer, electrooptic polarization plane switch, and deflector. The polarizer linearly polarizes radiation. The polarization switch turns by 90° the input beam direction of polarization by subjecting it to an electrical or magnetic field. It may be made of material with a linear electrooptic effect (e.g., crystals of $K_2H_2PO_4$, $NH_4H_2PO_4$, $LiNbO_3$, and many others) or a magneto-optic effect ($Y_3Fe_5O_{12}$, $CrBr_3$, etc.). The deflecting element performs spatial or angular separation of light beams with mutually orthogonal polarization.

The basic parameters of discrete electrooptic deflectors are as follows: number of deflection stages, number of light beam positions, working length of radiation wave, speed (i.e., time for switching from one resolution element to another), radiation attenuation factor, backlighting in switched-off resolution elements, and half-wave electrical voltage of radiation switching. Discrete deflectors may be both one and two coordinate.

The deflection of light beams by means of acoustic waves is based on the possibility of periodic spacial changing of medium density by means of acoustic waves, thus resulting in periodic spacial variation of the medium refraction coefficient. Acoustic waves form in the medium phase grid with a period equal to the wavelength. When the light beam passes through a medium with a sinusoidally varying refraction coefficient, light diffraction occurs. If the light beam falls at the running acoustic wave under a certain angle, only first-order Bragg diffraction is observed. With alteration of acoustic wave frequency, the direction of the deflected beam changes. Both liquid and solid isotropic and anisotropic materials may be used in Bragg deflectors. TeO_2, $PbMoO_4$, α-HgS, α-HIO_3, Ag_3AsSe_3, and many other materials feature high optoacoustic effectiveness. The following parameters characterize acoustooptic deflectors: beam deflection angle and resolution, speed, optical effectiveness characterized by the ratio of passed and falling radiation intensities, switching power, central (mean) acoustic wave frequency for which the Bragg condition holds, and the range of control frequency variations supporting light beam scanning.

Figure 5.17 gives comparative estimates of one-dimensional deflectors and their potentialities, provided they are built of the best materials: $LiNbO_3$ for electrooptic deflectors and α-HIO_3, $PbMoO_4$, and TeO_2 for acoustooptic deflectors. Comparison is carried out within the framework of the functional dependence of possible resolution on scanning time. It follows from the figure that there is a practical possibility of building two-dimensional multipositional deflectors with $\sim 10^4$ positions and switching frequency of up to 10 and more megacycles. The major parameters of commercially available deflectors are summed up in Table 5.3.

Multielement Electrically Controlled Space-Time Light Modulators

The effectiveness of space-time light modulators in optical data processing systems primarily depends on the properties of its working material, which should satisfy the following requirements: η, maximal effectiveness of optical signal transformation:

$$\eta = \frac{I_{in}}{I_{out}}$$

where I_{in} and I_{out} are, respectively, input and output signal intensities under maximal optical contrast or modulation depth;

Table 5.2 Light Source Manufacturers and Product Guide

Specifications	Units	ALLIANCE TECH. IND. ATI 22393	ASEA HAFO 1A83	ASEA HAFO 1AX114	ASEA HAFO 1AX113	BELL NORTHERN BNR-40-3-10-3	BELL NORTHERN BNR-40-3-15-3	BELL NORTHERN BNR-40-3-30-3
Light source type	–	LED	LED	LED	LED	LED	LED	LED
Peak radiated power	mW	.05	10	0.3	1.0	1.0	1.5	3.0
Drive current	mA	100	100	100	100	150	150	150
Peak wavelength	nm	820	940	910	860	840	840	840
Spectral bandwidth-50%	nm	35	60	–	–	45	45	40

Property	Units							
Half-angle beam spread	degrees	–	–	20	20	–	–	27
Emitting area size	μm	–	–	110	110	500	500	75
Rise time	ns	7	300	–	–	4	7	14
Operating temperature	°C	-40/85	-40/90	-40/90	-40/90	-40/85	-40/85	-40/85
Packaging unit	–	–	–	–	–	XO-72	XO-72	XO-72
Semiconductor material	–	GaAlAs	GaAs	GaAs	GaAs	GaAs	GaAs	GaAs
Operating voltage	V	3.4	1.5	2.0	2.0	1.7	1.7	1.7
Bandwidth	MHz	–	–	30	30	150	88	44
Threshold current	mA	–	–	–	–	–	–	–
Pigtail attached	–	YES	NO	–	–	YES	YES	YES
Overall size (L X D)	nm	12 X 25	–	–	–	11 X 25	11 X 25	11 X 25
Price	dollars	–	–	–	–	–	–	–

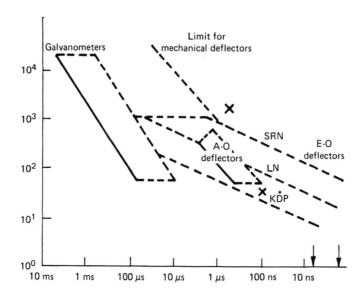

Figure 5.17. Comparison of potentialities of various types of deflectors; — present performance, -- projected performance, X digital E-O deflectors. (After D. Chen and I. Zook, *Proc. IEEE*, *63*(8), 1207, 1975.)

high sensitivity to the control signal, with the aim of reducing power consumption on switching of SLM elements; a memory effect (long-time memory with a special signal required to return the material to the original state, or relaxational memory), i.e., the ability to preserve the switched-on state during the time required to read out all the data array; no "fatigue" during the SLM lifetime; and a switching speed supporting the desired I/O speed.

The materials of electrically controlled space-time light modulators used for the transformation of electrical signals into light signals should feature essential nonlinearity (threshold property), which is essential for obtaining a large signal-to-noise ratio. This property is especially important in matrix addressing to elements by electrical voltage because it allows one to minimize the so-called cross-effect. Under matrix SLM addressing, which is also called rowwise addressing, control is exercised through $2\sqrt{N}$ channels, where N is the number of SLM elements.

Today, liquid crystal SLM are very popular and effective. Their array generation speeds, however, lie within the millisecond range. Faster SLM may be designed around PLZT ceramics. PLZT structure is a hot-pressed mixture of $PbZrO_3$, $PbTiO_3$, and lanthanum of composition $x/(y/z)$, where x is the atomic percentage of lanthanum and y/z is the ratio of $PbZrO_3$ to $PbTiO_3$.

By varying the mixture composition and conditions of hot pressing, one may obtain ceramics with various structural and optical properties, with or without long-time memory. These effects are due to the orientation of segnetoelectric domain polarization vectors in the electric field. As a result, the optical axes of crystal grains with pronounced double refraction are reoriented. There are some effects in ceramics that, in principle, may be used for spatial light modulation: a longitudinal electrooptic effect in stressed ceramics, a transverse electrooptic effect, a "boundary" effect, and an effect of light dissipation in large-grained ceramics. The dynamic possibilities of SLM based on electrooptic ceramics may be exemplified by a Sandia 256-element page composer operating at speeds up to 10^5 with a contrast ratio 1000:1.

SLM with individual addressing may be built around paraelectrical memoryless PLZT ceramics with higher optical switching speeds of the order of tens of nanoseconds.

Fast space-time modulators may be built around monocrystalline segnetoelectrics, which are bistable and suitable for SLM with memory. Their mechanism is based on the switching of spontaneous polarization, which is accompanied by essential changes in the optical properties of the crystal. Crystals of bismuth titanate ($Bi_4Ti_3O_{12}$) feature a small switching time of 1 μsec and a half-access voltage of 30 V. Large monocrystals of high optical quality may be obtained from gadolinium molybdate, $Gd_2(MoO_4)_3$, but their switching times are about 0.1 to 1 msec, which is much greater than those of bismuth titanate.

Ferromagnetic materials may be expected to provide a good basis for the design of high-speed spatial light modulators that are matrix addressable by electrical voltage. Their true switching threshold enables the design of devices with over 100 X 100 resolvable elements. There are published reports of SLM designs made of orthoferrites, such as $DyFeO_3$, $YFeO_3$, and $YFe_{1-x}Gd_xO_3$, that feature the following parameters: 100 X 100 elements; maximal contrast, 400:1; and an optical effectiveness for the red line of a helium-neon laser of over 15%. Writing is done by local magnetic fields generated by current loops, one data bit requiring 10^{-8} sec under a current of 1 A at most and a current loop

Table 5.3 Basic Parameters of Commercially Available Deflectors

Manufacturer	Model	Working wavelength, μm	Allowable positions (by Rayleigh criterion)	Working frequency range, Mcps
SORO	D 250K	0.44-0.7	800	150-300
		0.44-0.7	1000	150-300
ISOMET	LD-400-X	0.514	650	75-125
	LD-400-X	0.514	400 X 400	75-125
	LD-401-1X	0.44	320	85-125
	LD-401-2X	0.488-0.514	400	75-125
	LD-401-5X	0.6328	400	75-125
	LD-401-7X	1.06	180	35-55
ZENITH	D 70R	0.6328	400	50-90
	D 150R	0.6328	1000	100-200

diameter of 100 to 300 μm. Erasures and rewriting may be repeated arbitrarily; the storage time is also not limited.

It is hoped that new materials and semiconductor injection lasers will enable one to overcome difficulties due to the appreciable absorption of ferromagnetics in the visible spectrum and to the high currents required for magnetic flux reversal.

Space-time light modulators based on multichannel acoustic modulators are also feasible. The light beam is modulated and deflected by means of light diffraction on an acoustic wave excited by piezocrystals. A data array may be generated by means of a great number of individually addressable cells whose number corresponds to the number of modulation channels. The time sequence of m electrical pulses in n channels is transformed by the modulator into a running picture consisting of m X n elements.

Effective-ness, %	Control-ling power, W	Soundguide material	Optical aperture, mm	Switching time, μsec	Notes
>50	-	$PbMoO_4$	20	5.5	-
>50	-	$PbMoO_4$	24	7.5	-
>50	3.5	TeO_2	8 X 7	13	
>50	3.5	TeO_2	φ 6.6	10.7	2-D de-flection
-	1.5	TeO_2	φ 6.6	10.7	-
-	2	TeO_2	φ 6.6	10.7	-
-	3	TeO_2	φ 6.6	10.7	-
-	4	TeO_2	φ 7.0	12	-
Center 70, edges 60	3.5	Glass	38.6 X 2	10	-
Center 50, edges 40	9	-	41 X 2	10	-

For parallel reading of a two-dimensional array, the pulse duration of the reading beam should be sufficiently short (about tens of nanoseconds) in order that a group of successively excited acoustic signals with one carrying frequency might be reagarded as a stationary diffraction grid. An amplitude spatial modulator has been designed with the following characteristics: number of channels $n = 34$, number of elements in each channel $m = 128$, power consumption per channel ~ 0.5 W, and length of elementary acoustic beam 0.127 mm for modulator aperture 33.6 X 21.7 mm^2. Modulator capacity is estimated as 1 Gbps. Multichannel acoustic modulators require powerful pulse sources because the acoustic picture may be regarded as stationary only during a period of several nanoseconds. The major characteristics of some electrically controlled space-time light modulators are tabulated in Table 5.4.

Table 5.4 Basic Characteristics of Some Electrically Controlled SLM

Characteristics	Segnetoceramic SLM		Acoustooptic SLM		Magnetooptic SLM
	PLZT 7/65/35 grain, 2 μm	PLZT 9/65/35 grain, 2 μm	PbMoO$_4$	TeO$_2$	YFeO$_3$
Modulating medium	PLZT 7/65/35 grain, 2 μm	PLZT 9/65/35 grain, 2 μm	PbMoO$_4$	TeO$_2$	YFeO$_3$
Modulating effect	Double refraction change	Double refraction change	Refractive index change	Refractive index change	Faraday effect
Addressing	Matrix	Individual	Individual	Individual	Matrix
Working aperture, cm	3.2 X 3.2	3.0 X 0.5	3.4 X 2.2	8.6 X 75	2 X 2
Supply voltage, V	100-200	300	-	-	Current 1 A
Complete resolution	128 X 120	128 X 1	34 X 128	100	100 X 100
Element switching time, sec	-	10^{-7}	5.10^{-6}	10^{-6}	10^{-8}
Array generation time, sec	-	5.10^{-7}	5.10^{-6}	10^{-6}	10^{-5}
Cycle time, sec	-	5.10^{-7}	5.10^{-6}	10^{-6}	10^{-5}
Memory	Permanent	No	No	No	Permanent
Maximal contrast	10:1	100:1	-	30:1	350:1
Lifetime	Up to 10^7-10^{10} cycles	10^{11}	Possibly unlimited	Possibly unlimited	Possibly unlimited

Electrically controlled SLM input data arriving from the computer memory in electrical form into the optical processing path. Their major task is to quickly generate optical files, i.e., to prepare data for parallel optical systems. Thus, they serve as interfaces between electronic and optical channels of the computer system. As indicated above, their throughput may run into 10^{12} bps and more, but currently existing modulators exhibit a throughput of about 10^9 bps. Data input devices based on lines of memoryless modulators with individual addressing have good speed and may be addressed in one cycle; therefore, a line of 100 elements may have throughput of 10^9 to 10^{10} bps.

Optically Controlled Digital Space-Light Modulators

The best evolved are optically controlled spatial light modulators for analog coherent optical real-time data processing. There is a large choice of materials for optically controlled spatial light modulators and corresponding devices. For their description the reader is referred to published surveys. The requirements for optically controlled SLM greatly depend on the particular application, and possibly none of the existing designs may be regarded as general purpose.

The optically controlled digital SLM are intended for processing in one cycle of flows of falling digital data arrayed in two dimensions in the form of binary pictures, tables, etc.

The digital optically controlled SLM is a central logic element in optical computer engineering. It performs a double function of logical transformation of optical pictures and restoration of the luminous energy of the resulting picture up to the standard level at the expense of the power supply. These functions are essential because any discrete picture transformation entails significant luminous losses. In the absence of such elements, the possibilities of optical processors are tangibly limited, because without restoration of picture intensity to the standard level, only systems with few (two or three) processing stages are feasible, and computations in which the result should be fed back from the processor to the memory and again to the input become impossible.

A popular method of designing OCSLM is based on a multi-layered structure consisting of photoresistor and electrooptic material (see Fig. 5.18). OCSLM consists of layers of photoresistor, electrooptic material, and an optical decoupling layer. On both sides of the structure, transparent electrodes are deposited through which control voltage is applied to the SLM.

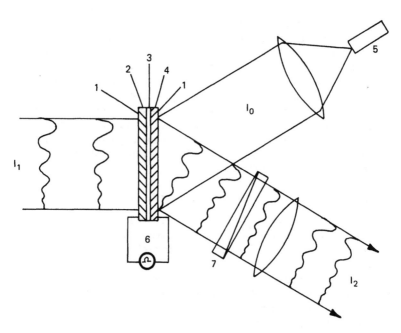

Figure 5.18. Scheme of optically controlled SLM with a uniform modulating layer: I_0, reading wave; I_1, input picture control wave, and I_2, output picture. (1) Transparent electrodes, (2) photoconducting layer, (3) optical decoupling layer, (4) electro-optic light-modulating layer, (5) reading wave source, (6) voltage pulse generator, and (7) polarization analyzer.

The optical decoupling layer is made as a matrix of mirror-reflecting metallic cells separated by a nontransparent resistive layer. Along with the absorbing layer, a mask may be used between reflecting cells deposited on the surface of a transparent electrode, which prevents light from falling between the mirror-reflecting cells (Fig. 5.19). The dielectric mirror and absorbing layers allow one to significantly reduce the influence of reading radiation on photoresistor excitation and enable data read and write by radiation of the same wavelength.

Each cell of OCSLM operates as a voltage divider with a photoresistor in one of its arms. The control wave causes redistribution of the voltage applied to the structure. As a result, the reading wave, after having been reflected from SLM and

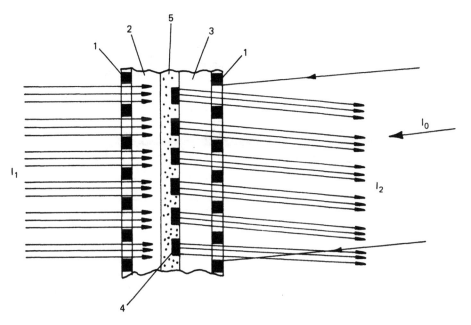

Figure 5.19. Structure of an optically controlled reflecting SLM with an optical decoupling layer: I_0, reading wave; I_1, control wave; and I_2, reflected wave with output picture. (1) Masking electrodes with transparent windows, (2) photoconductor, (3) electrooptical material, (4) mirrors, and (5) absorbing resistive layer.

passed twice through the electrooptical layer, is modulated in amplitude or phase in lighted cells. When the pulse is completed, the structure relaxes to the initial state.

In planar SLM designs, the reflected wave may be modulated by means of such effects as a longitudinal electrooptic effect on dielectric crystals, such as ADP, KDP, and KTN, electric absorption due to the Franz-Keldysh effect in wide-zone semiconductors, such as GaAs, CdS, and GaP, dynamic scattering in liquid crystals under the electrical field, etc.

Optically controlled SLM with a uniform photosensitive layer are the best studied and today are more popular than other SLM types. Although their performance (sensitivity, speed, and picture contrast) is steadily improving, their use as a basic build-

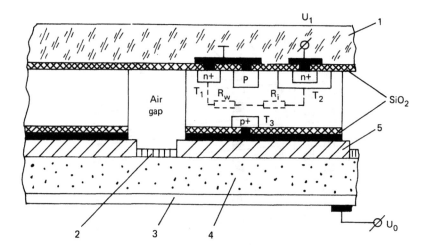

Figure 5.20. Scheme of digital optically controlled SLM of latrix type: (1) glass, (2) nontransparent dielectric, (3) transparent electrode, (4) electrooptical material, and (5) metal.

ing block for universal optical digital computer engineering is questionable. First, for implementation of memory in these structures, one usually employs for light modulation electrooptic materials such as magnetoelectrics featuring significant residual polarization. This means that, for a material thickness of about 50 μm, the repolarization energy liberated in a write-erasure cycle within the electrooptic layer is about 10^{-4} J/cm^2 and more. The allowable energy liberation for SLM with a speed of about 10^{-6} sec or more and with normal heat sink about 1 W/cm^2 is about 10^{-6} J/cm^2. Second, the speed of uniform layer SLM are limited, and cycle frequencies in this case cannot be more than 10^4 to 10^5 cps. This is due to the impossibility of obtaining high photocurrent gain in a uniform photosensitive layer.

Taylor and Kosonocky proposed an optically controlled digital SLM consisting of photoreceivers, transistor control elements, and a light gate as early as 1972.

Figure 5.20 shows the circuit diagram of an element consisting of a cell of a silicon control-integrated matrix connected by a mirror metallic pad to a layer of electrooptic material. The integrated matrix is fabricated on a glass wafer transparent to control light.

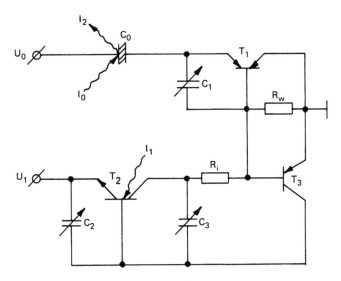

Figure 5.21. Equivalent electric circuit of digital SLM with internal electronic logic: I_1, input picture control wave; I_0, reading wave, and I_2, output wave.

An equivalent electrical diagram of the cell is depicted in Fig. 5.21. The electrooptic layer covered by a transparent electrode is common to all the cells. The electronic control portion of the SLM cell consists of a key T_1 through which the modulating layer capacitance c_0 is charged and discharged. The state of the key T_1 is defined by the photothyristor on transistors T_2 and T_3. The circuit operates in the following manner. To give rise to voltage in the layer of electrooptic material, key T_1 is opened by an electrical pulse sent to its base by the thyristor, which at this stage is controlled electrically. A positive voltage pulse of amplitude u_0 is applied to the transparent electrode, and the current charges the modulating layer capacitance. Then the key is cut off, and the supply voltage pulse is switched off. Now, voltage u_0 is distributed between capacitance C_0 and capacitance of the backward biased p-n junction of transistor T_1, the voltage on the modulating layer being 0.8 to $0.9u_0$. However, only non-lighted latrix cells preserve the charge on C_0. If a light pulse I_1 falls on a cell, the photothyristor opens key T_1 through which C_0 is discharged. Thus, writing results in the polarizing modulation of reading light owing to an electrooptic effect in the electrooptic layer.

Such control of voltage in the elementary latrix cell radically differs from control in uniform photolayer structures operating like a resistor-capacitor (RC) divider. In the case under consideration, the structure operates as a capacitor divider, thus bringing about advantages in speed and power consumption. An additional advantage is the ability of the cell to keep data during the time of leakage through a cutoff p-n junction, i.e., during 10^{-2} sec or 10^3 to 10^4 clock cycles. In each cycle, voltage reduction in the electrooptic material by 10^{-3} to $10^{-4}u_0$ should be offset. That only a portion of the voltage should be restored increases SLM cost effectiveness. In contrast to conventional static storage, dynamic storage enables employment of hysteresisless electrooptic materials, i.e., magnetoelectrics at temperatures above the Curie point. As a result, the density of power dissipated in a write/erasure cycle drops to 10^{-7} J/cm^2. The value of control voltage on the modulator is defined by the p-n junction breakdown voltage of transistor T_1 and may be about 100 V.

Supposedly, the general-purpose SLM will have the following characteristics:

Cycle frequency, 1 to 10 Mcps
Number of cells, 10^3 to 10^4
Brightness gain (i.e., ratio of power of a single element at the transparency output to the same power at its input), >30
Sensitivity, 10^{-13} to 10^{-14} J per element, which is due to high internal amplification of photocurrent

The digital optically controlled space-light modulator with the above or even better performance is the basic building block of optical digital computer engineering.

Materials for Optical Memory

Optical memory has a high density of stored data and fast access to the arbitrary memory cell. Together with the parallel fetch of two-dimensional data arrays and input into optical arithmetic units, this accounts for the great interest in optical memory.

Materials for optical memory should satisfy a number of requirements common to any computing device and specific to optical data storage. From the viewpoint of computer engineering, materials for optical storage should maintain their parameters during a long time interval, usually several years, but the environment should not include freezing to cryogenic levels or

vacuum. Depending on memory type, materials should be reversible to some degree. Memory storing constants and instructions
requires only one data writing, but the number of write/erasure
cycles in main memory is tremendous, and the optical properties
of its medium should be permanent.

Optical data writing imposes the following requirements: low
writing energy; high sensitivity to radiation of popular lasers,
such as helium-neon, argon, helium-cadmium, and gallium ar-
senide-based semiconductor lasers; and high material resolution.

Photographic plates and films are the most popular and best
developed materials for read-only optical memories because they
feature high resolution (over 2000 lines/mm) and sensitivity
(10^{-5} to 10^{-4} J/cm^2).

High-quality holograms may be obtained by means of dichro-
mated gelatin, $(NH_4)_2 Cr_2O_7$, or $K_2Cr_2O_7$. Optical properties of
such holograms are approaching the ideal. Both reflection and
transmission holograms may be written on this material, with high
diffraction effectiveness combined with low noise level. The material sensitivity is about 10^{-3} J/cm^2, and resolution is about 4000
lines/mm.

Some physical effects and phenomena may be used for writing
data: for example, thermomagnetic and thermal writing and photoinduced processes. In thermomagnetic writing, laser radiation
heats material up to the temperature of the magnetic flux reversal.
Reading is based on the magnetooptic Kerr and Faraday effects.
Studies have been carried out on such materials as MnBi, EuO,
GdIG, YIG:Gd, and MnAs.

Studies are in progress on reversible optical data recording on
films of amorphous semiconductors of various compositions. For
example, alloys of $Ge_{15}Te_{85}$ under heating have a phase transition
into the crystalline state that is accompanied by a change in the
film optical characteristics. The required light pulse energy is
close to that required for writing on MnBi, i.e., about 10^{-2} J/cm^2.

Under certain temperatures, vanadium oxides undergo a phase
transition from a low conductivity state into that of very high
metallic conductivity. The phase transition of vanadium dioxide
(VO_2) occurs at temperatures about 70°C. Low and high conductivity states have different coefficients of reflection and transmission.

In photoinduced processes, the medium directly interacts with
photons. Photochromatic materials change their absorption spectra
when exposed to short-wave irradiation. As a result, their transmission capability at the reading wavelength changes. Information

thus written may be erased by long-wave irradiation or by heating. The photochromatic materials have high resolution (over 3000 lines/mm), do not require development or fixation, and may store a great many holograms written under different angles. At the same time, photochromatic materials have low sensitivity, 10^{-1} to 10^{-2} J/cm^2, require two wavelengths for data write and erasure, and have low hologram diffraction effectiveness and low temperature stability.

When illuminated by an intensive light field, some segnetoelectrics change their refraction coefficient in the lighted area. This effect found in $LiNbO_3$, $LiTaO_3$, $BaTiO_3$, $Bi_4Ti_3O_{12}$, $BaCaNaNbO_3$, etc., may be used for holographic data recording. Information may be erased in crystals by heating or illumination. Among the merits of magnetoelectric crystals one may mention high resolution (over 4000 lines/mm) and high diffraction effectiveness; also, no "fatigue" is observed after numerous write/erasure cycles. Rather, the high write energy of about 10^{-1} to 10^{-2} J/cm^2 is a disadvantage of these crystals.

Reasoning from the fact that the sensitivity of materials for reversible optical holographic memory has to be about 10^{-11} to 10^{-12} J for recording a data bit, i.e., higher than 10^{-5} to 10^{-6} J/cm^2, and that the write/erasure cycle should be 1 μsec or more, one may conclude that none of the existing or studied materials has all the properties required for the design of optical main memory with parallel writing of data arrays with 10^3 to 10^4 bits.

As appropriate reversible media are lacking, files may be stored in an optoelectronic processor by general-purpose digital SLM-like "latrix," which differs from the above SLM by having two equivalent control emitters T_2 instead of one. Both emitters may be connected to the corresponding emitters of other cells by means of common buses operating as orthogonally addressable electrodes. In this case, the device may be used as read/write memory, as well as an optically and electrically controlled space-time light modulator.

Optical Elements

Apart from logic transformations, optical transformations of digital pictures also play an important role in optical computer engineering. These transformations occur in the transmission of digital pictures from SLM to SLM and perform the function of optical path switching. In contrast to logical transformations,

optical elements do not need computer time because the time of
light propagation through optical transformers is much less than
that of SLM response.

Some simple transformations over digital pictures may be done
by means of conventional cylindrical or raster optics. The logical
addition of row or column elements is done, for instance, by a
cylindrical lens. The multiplication of numerical column or row
elements is done by a cylindrical lens raster. Transposition,
i.e., substitution of columns for rows, may be also done by a
cylindrical raster, where the axes of cylindrical lenses are
oriented along the main diagonals of digital cells.

Wide possibilities of optical picture transformations are opened
by the advent of fiber optics. Superposition and branching of
numerical pictures may be done by the device shown in Fig. 5.22.
Fiber optical elements may also perform transposition and shift of
a numerical picture, etc.

Thus, optical operations over digital pictures are performed
by traditional elements, and their design does not give rise to
significant technological problems for the optical industry.

5.4 DESIGN OF ELEMENTS FOR OPTOELECTRONIC PROCESSORS BY METHODS OF INTEGRATED OPTICS

Of prime importance for the implementation of high-throughput
optoelectronic processors are effective and fast multipositional
deflectors and digital space-time light modulators.

Section 5.3 deals with the implementation of space-time modulators and deflectors on the basis of electrooptic, acoustooptic,
and magnetooptic effects in volume media. Being rather weak,
these effects require strong electric, magnetic, and acoustic
fields for light modulation and deflection. Hence, energy stored
by the field necessarily should be high; the consumed power dissipated as heat is proportional to the stored energy. This leads
to a conclusion that a technology is required to enable the design of space-time light modulators and deflectors with minimal
power consumption.

Designs of waveguide optical modulators and other integrated
optics devices are based on film optical waveguides. Their properties and design techniques are described elsewhere and may be
summarized as follows.

An optical waveguide is a thin film of thickness h and with refractive index n_1 deposited on a wafer. On the reverse side,

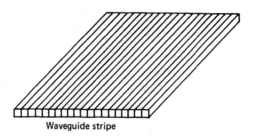

Waveguide stripe

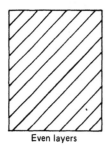

Even layers

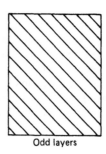

Odd layers

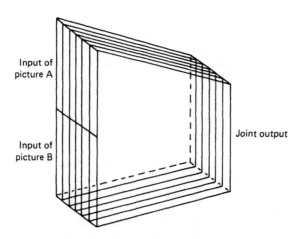

Figure 5.22. Fiber optics-based device matching two pictures. (After D. H. Shaefer and J. P. Strong, *The Computer, Proc. IEEE,* 65(4), 129-138, 1977.)

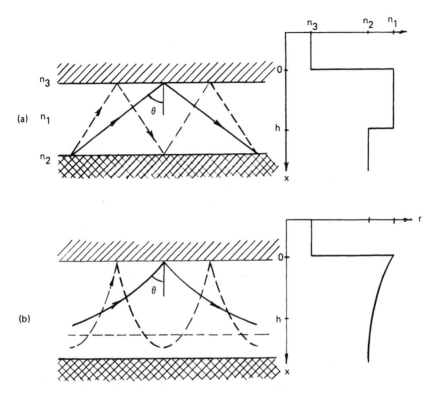

Figure 5.23. Refractive index distribution and light propagation in optical waveguides. (a) Stepwise change of refractive index. (b) Continuous change of refractive index. Solid and dotted lines refer to lower and higher modes, respectively.

there is a covering layer (or air), as shown in Fig. 5.23. Refractive indices of the wafer n_2 and cover n_3 are smaller than those of the film:

$$n_1 > n_2 \geq n_3$$

As shown in Fig. 5.23, variations of the refractive index in the waveguide may be stepwise or may have a gradient. In any case, the condition $n_1 > n_2 \geq n_3$ assures radiation waveguide propagation in the film owing to the complete internal reflection from interfaces. Differences of refractive indices in this case should be 10^{-3} to 10^{-2}.

Physically, waveguide light propagation means that light in the film follows a zigzag path. Field amplitude in the direction of propagation along the film z is defined by exp $(-j\beta z)$, where the propagation constant β is related to the angle θ by

$$\beta = kn_1 \sin \theta$$

where $k = 2\pi/\lambda$ is the wave number. TE and TM modes may propagate in the waveguide, polarized along or across the film plane, respectively. The number of oriented modes is discrete and grows with waveguide thickness h. In addition, in the waveguide structure, modes may occur whose angle of incidence with the interface is smaller than that of complete internal reflection. Such modes leave the waveguide in the direction of the wafer or cover, and they are referred to as radiation modes.

Since optical waveguide thickness is typically several micrometers, special devices are required for radiation input. Two kinds of such devices, prism and grid, are shown in Fig. 5.24. In the former case (Fig. 5.24a), a prism with larger refractive index $n_p \gtrsim n_1$ is mounted in the vicinity of the film waveguide surface (usually at a distance of about 1000 Å), to be able to interact with the waveguide mode field. To support effective radiation input, the following condition of phase matching should be met, i.e.,

$$\beta = kn_p \sin \theta_p$$

where θ_p is the prism input angle, normal to the film.

For a grating input device (Fig. 5.24b), a periodic structure (e.g., diffraction grating) is made in the waveguide. The general characteristics of prism and grating input devices are sufficiently similar. In order to provide uniformity of connection, i.e., a constant gap between the film and prism or a constant grating step, the maximal theoretical input effectiveness is estimated as 80% for a light beam with uniform or Gaussian cross section. If the gap between the prism and film narrows to match input beam and waveguide output field, or if grid thickness and profile are appropriately calculated, the input effectiveness may reach almost 100%. Optical waveguides may be fabricated on the wafer surface by etching, ion implantation, or other means, and also by depositing on its surface films of other substances.

At present, the most popular materials for electrooptic and acoustooptic modulators are lithium niobate and tantalate. In these materials, a waveguide layer is fabricated with a gradient of refraction index. This may be done by heating the crystal up to 900°C. In doing so, LiO_2 diffuses from the crystal (back diffusion); thus, the surface layer becomes lithium depleted, which results in a refractive index greater than usual. Another

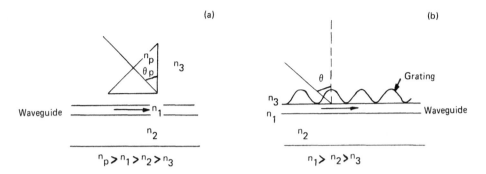

Figure 5.24. (a) Prism coupler. (b) Input of light into thin-film waveguide by means of grating.

method is vaporizing a metal (most often, titanium) on a crystal surface. When heated to 950°C, the metal diffuses into the crystal, causing changes in both refractive indices. This method is fairly close to the microelectronic planar technology and enables fabrication of various waveguide structures.

Gallium arsenide is another popular material offering promise for integrated optical devices for laser radiation control because it enables the design of film lasers, modulators, photodiodes, and electronic control circuits, i.e., the integration of various optical components on the same wafer.

Waveguides are being designed to be fabricated on silicon wafers by means of a technology compatible with that of silicon integrated circuits (Fig. 5.25).

Among the numerous integrated optical modulators, a directional coupler waveguide switch may be singled out (Fig. 5.26). Two similar strip waveguides made of electrooptic materials are deposited on a common wafer so that they are parallel and isolated by gap g. The length of area where they are parallel is ℓ. Outside this area the distance between the waveguides is much greater than g. The disposition of electrodes is shown in the figure. If waveguide propagation constants are equal ($\beta_1 = \beta_2$) and waveguides are similar, the luminous flux from waveguide 1 is pumped into waveguide 2, and vice versa, provided that the gap g is so small as to assure overlapping of decaying fields of both waveguides. If ℓ is very large and all the light initially was in waveguide 1, after a certain distance all the luminous flux passes to waveguide 2; then, after the same distance, the

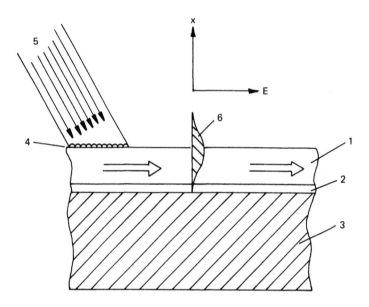

Figure 5.25. Integrated optical waveguide on silicon wafer: (1) transparent waveguide layer, (2) layer of SiO_2, (3) silicon layer, (4) diffraction grating for input of light into waveguide, (5) falling light wave, and (6) field amplitude distribution in waveguide.

light passes into the first waveguide, and so on. \The distance at which the light is completely pumped from waveguide 1 to waveguide 2 with matched phases is referred to as critical connection length:

$$L_0 = \frac{\pi}{2\kappa}$$

where κ is known as the connection constant. Without voltage $\beta_1 = \beta_2$, and if $L = \ell$, all the light is pumped from waveguide 1 into waveguide 2. With field applied, phase matching is violated ($\beta_1 \neq \beta_2$) and the amount of light pumped from one waveguide to another changes. If

$$\left(\frac{\beta_1 - \beta_2}{2\kappa}\right)^2 \gg 1$$

the amount of transmitted energy drops dramatically.

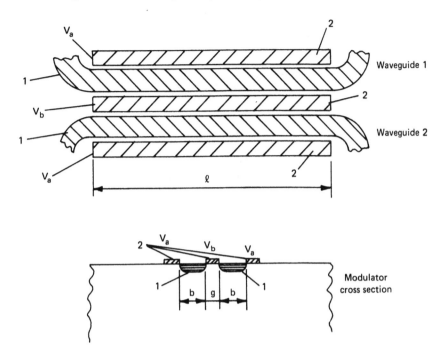

Figure 5.26. Integrated optical modulator based on coupled strip waveguides: g, distance between waveguides; ℓ, waveguide width; (1) waveguides, and (2) electrodes. [After *Integrated Optics*, Second Edition (T. Tamir, ed.), Springer-Verlag, New York, 1979.]

Thus, if no field is applied, the light passing through one waveguide goes to another; if there is a field, the light stays in the same waveguide. The system of connected waveguides may be regarded as an elementary logic gate (Fig. 5.27 a, b). Let V_0 be a potential under which there is no pumping from one waveguide to another, and radiation be put into the first waveguide and sensed also at its output. Then, if potentials at two signal electrodes V_a and V_b are $V_a = V_b = 0$, or if $V_a = V_b = V_0$, there is no light at the output of waveguide 1; if $V_a = 0$ and $V_b = V_0$, or $V_a = V_0$ and $V_b = 0$, the light stays in the initially excited waveguide, i.e., mod 2 addition ($a \oplus b$) is implemented, the function $\overline{a \oplus b}$ being realized at the output of waveguide 2.

Complete switching off of lithium niobate at $g = 2$ μm was estimated to occur at $V_0 = 2$ V, the critical connection length being

(a)

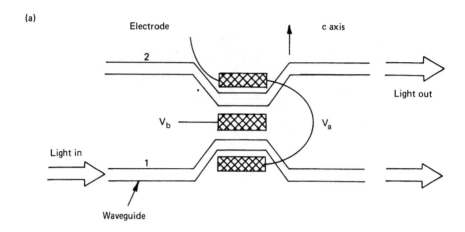

(b)

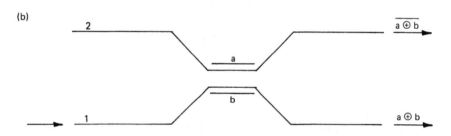

Figure 5.27. (a) Integrated optical switch based on coupled waveguides: (1) waveguide into which light is input and (2) passive
waveguide at whose output light may appear. (b) Logic diagram
of coupled waveguide-based integrated optical switch. (After
H. T. Taylor, *Appl. Optics*, 17(10), 1493, 1978.)

L = 4 mm. For g = 2 μm and L = 4 mm, power consumption is
about 2.10^{-5} W/Mcps. At a switching frequency 10 Mcps, power
consumption is about 2.10^{-4} W per element and switching energy
is 20 pJ.

This example demonstrates that the waveguides may serve as
a basis for electrooptic switching devices with outstanding characteristics. By comparison, the up-to-date high-speed ECL and
Schottky TTL electronic logic circuits require switching energy
within the range 50 to 100 pJ.

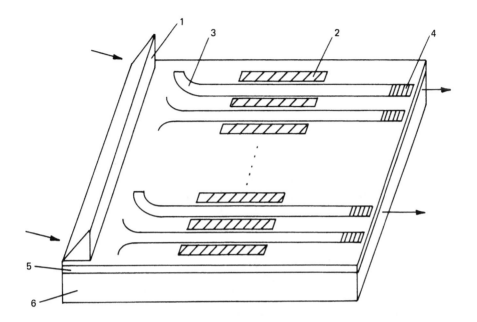

Figure 5.28. Multielement linear SLM based on coupled waveguides:
(1) prism input element, (2) electrodes, (3) strip waveguides, (4)
diffraction grating for light input, (5) thin-film waveguide, and
(6) wafer.

It is clear from the discussion above that, for optoelectronic
processors, linear and matrix space-time light modulators and
multipositional deflectors with outstanding parameters may be
designed around integrated optical structures.

The diagram of a linear space-time light modulator relying on
connected waveguides is shown in Fig. 5.28. Linear modulators
with 32, 64, or 128 elements input multipositional numbers from
the electronic main memory into the optoelectronic processor.
Input numbers are automatically represented in paraphase code,
which is the advantage of the linear modulator on connected
strip waveguides.

Data in optoelectronic processors are handled by space-time
light modulators with optical input and output. Integrated optics enables the design of optically controlled SLM with significantly higher speed, lower power dissipation, and smaller size.

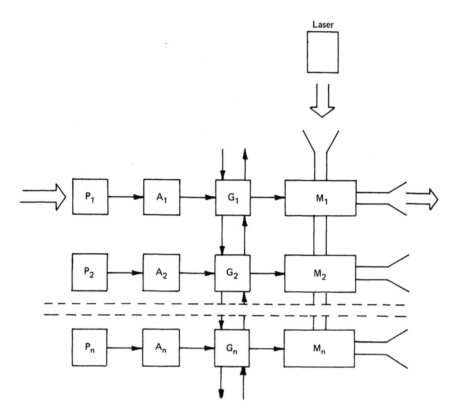

Figure 5.29. Functional diagram of row of integrated optical multi-channel SLM: P_i, photodiodes; A_i, amplifiers; G_i, logic gates, and M_i, waveguide branchers with modulators.

At the same time, controlled SLM on optical integrated circuits are technologically much simpler.

Figure 5.29 depicts the functional diagram of a controlled row of SLM as implemented by means of integrated optics. Control light signals falling on photodiodes P_i of row cells are electronically amplified and fed into electronic logic elements, which may be interconnected in some way. Reading light from an external source through waveguide couplers arrives at the optical modulators and then at the radiation output elements. A uniform distribution of reading light power with respect to row elements may be done in various ways; typical examples are shown in Fig. 5.30a, b. In

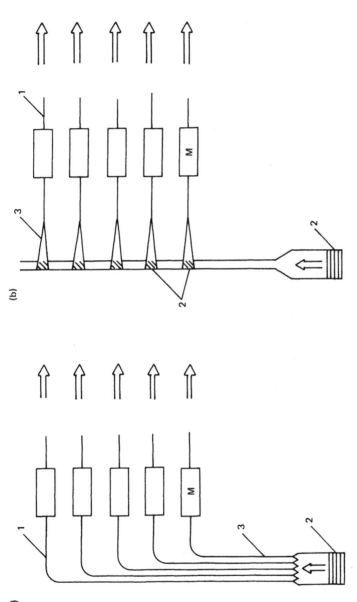

Figure 5.30. Schemes of waveguide light tracing by means of (a) focon branchers, and (b) diffraction gratings: (1) stripe waveguide, (2) g, diffraction grating, (3) planar focon, and (M) modulator.

the first version, the input wave coming through a diffraction grating into a wide (about 1 mm) waveguide is branched by planar focons into separate single-mode waveguides. Distribution is done by the rotation of light channels. In the second version, partial couplers to the main planar waveguides are used, implemented by means of Bragg gratings and reflecting the light wave under some angle to the main waveguide. To ensure uniform distribution of light power between the row elements, the grid effectiveness along the main line should be variable. Branched waves arrive at cell waveguide channels through focon inputs.

The design of digital SLM as a whole, evidently, depends on the method of connecting rows into pages. In integrated SLM, all the rows are placed in the wafer plane (Fig. 5.31a). On one side of the wafer, photodiodes are made; on the other side, integrated optical waveguides and modulators are fabricated. Reading radiation is input in the main waveguide and branched into row waveguides, thus reaching separate SLM cells. Another method of radiation input is possible in which each SLM cell has its own grating radiation input (Fig. 5.31b). On this case, the logic interaction of two pictures becomes possible, one being input from the side of photodiodes and another from the side of integrated optical waveguides.

In the hybrid version of two-dimensional SLM, each row is implemented on a separate wafer, a page being formed by assembling a stack of row wafers. Each SLM row has a silicon integrated circuit and a transparent plate carrying an optical integrated circuit. The optical integrated circuit contains waveguide modulators and radiation input and output elements; the silicon integrated circuit has a row of photodiodes, amplifiers, and logic elements. An example of an SLM row fabricated by hybrid technology is shown in Fig. 5.32. A two-dimensional SLM consisting of a set of rows is shown in Fig. 5.33. Radiation may be conveniently output through waveguide butt ends by means of diffraction grating, which deflects the radiation under a small angle into the wafer (Fig. 5.34). In the same figure a technique is shown of coupling two SLM by means of mirror surfaces of skewed wafer butt ends. Sometimes this design of input and output elements allows one to pass the output picture of one SLM to the input of another without intermediate optical elements.

Let us estimate feasible parameters of an integrated optical SLM. The element switching energy and control voltage are important for attaining high speed. Modulator switching energy is characterized by the ratio of control power P to the bandwidth of

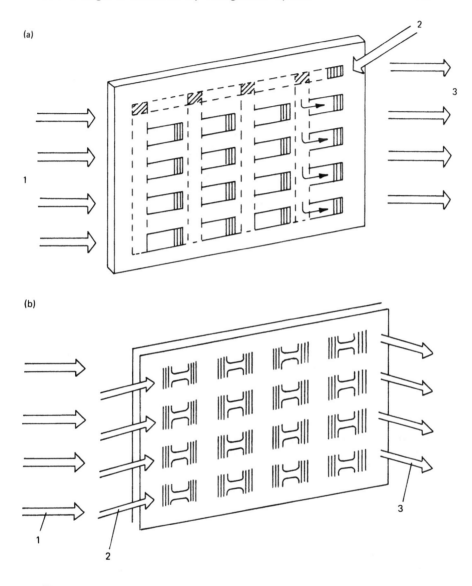

Figure 5.31. (a) A version of a two-dimensional integrated optical SLM with a common input of reading light: (1) input picture, (2) laser beam, and (3) output picture. (b) A two-dimensional integrated optical SLM based on coupled waveguides: (1) operator fields, (2) input variables, and (3) output variables.

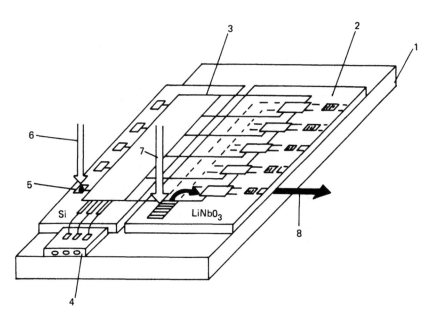

Figure 5.32. An example of hybrid integrated optical SLM: (1) plate, (2) optical IC, (3) semiconductor control circuit, (4) electrical plug, (5) photodiode, (6) input picture, (7) laser beam, and (8) output picture.

modulation frequencies Δf under a modulation coefficient of about 1. Strip waveguide modulators made of gallium arsenide and lithium niobate have $P/\Delta f = 10$ µW/Mcps; with further technological progress this value may be reduced to 1 µW/Mcps, i.e., 10^{-12} J for switching one element under control voltage 10 V, at most. Integrated optical modulators are compatible with photodiodes and logical elements of silicon integrated circuits both in switching energy (10^{-11} to 10^{-12} J) and control voltage (3 to 10 V); thus, they can operate at frequencies of tens and even hundreds of megacycles, which is comparable with the speeds of modern computers.

SLM throughput for given power of reading laser P_i and energy sensitivity of photodiode E_{ph} (J/bit) is defined as follows:

$$Nf_T = 10^{-0.1G} \frac{P_i}{E_{ph}}$$

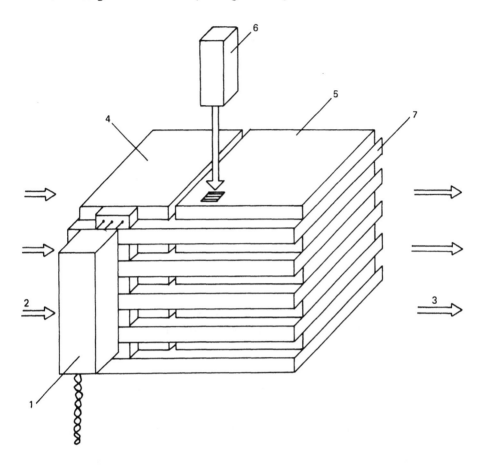

Figure 5.33. An example of two-dimensional optically controlled SLM consisting of a set of rows: (1) electrical plug, (2) input picture, (3) output picture, (4) semiconductor control circuit, (5) optical IC, (6) laser, and (7) transparent plate.

where N is the number of picture elements, f_T is clock frequency, G is light power gain offsetting losses for input, output, and branching of output signal per picture. Assuming that, for the radiation output technique shown in Fig. 5.31b, input and output losses of one cell are G = 10 db, the number of elements N = 64 X 64, the sensitivity of the silicon photodiode with amplifier E_{ph} =

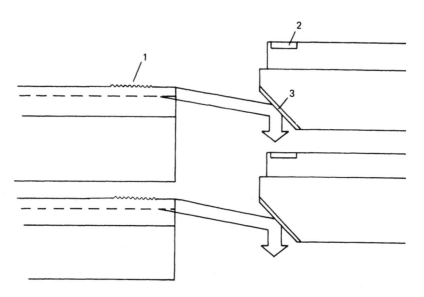

Figure 5.34. Picture transmission between SLM: (1) diffraction grating, (2) photodiode, and (3) mirror face of wafer.

10^{-14} J, and reading laser power P_L = 20 mW, obtain an SLM clock frequency of about 50 Mcps or throughput about 10^{11} bps, which is close to estimate (5.1) of Sec. 5.2.

Integrated optical methods may prove to be effective for read-only holographic memories storing programs, instructions, task execution algorithms, etc., in the form of two-dimensional binary pictures. Figure 5.35 shows a scheme of writing waveguide holograms. First, a thin layer of light-sensitive medium is put on a thin-film waveguide. The reference beam is input through the wafer by means of a prism. A Fourier image of written SLM falls on the light-sensitive layer, also from the same side. If As_2S_3 is used as the light-sensitive medium, holograms may be written, for example, by the line 5145 Å of argon laser. Hologram restoration may be done, for instance, by wavelength 6328 Å of a helium-neon laser. Aberrationless hologram restoration requires matching the propagation constants of reference and restoration waves. Experimentally, waveguide holograms were written on 3000 to 8000 Å thick films of As_2S_3 hologram diameter about 1 mm^2.

A high diffraction effectiveness is a distinguishing feature of waveguide holograms. The restoration by waveguide TE mode

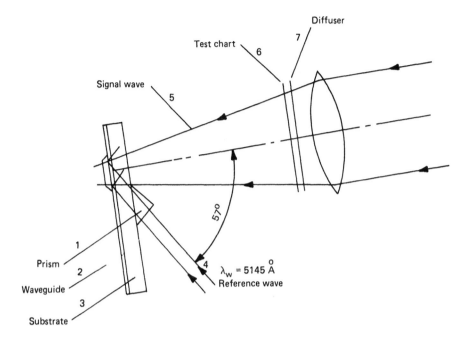

Figure 5.35. Scheme of waveguide hologram writing: (1) prism, (2) waveguide, (3) wafer, (4) reference wave, (5) object wave, (6) transparency, and (7) diffusor.

has given a complete diffraction effectiveness of 92%, which is in good agreement with the predicted theoretical result.

One hologram of 1 mm² accommodates 10^3 to 10^4 bits, and a 5 X 5 cm plate accommodates 1000 holograms. Light may be switched from one hologram to another by connected strip waveguides or controlled Bragg gratings. Such an optical memory is superior to its analog constructed of three-dimensional elements.

Integrated optics may be used to design fast multipositional deflectors in which the beam is deflected, for example, by light and acoustic waves interacting in thin-film waveguide structures. If a light beam of diameter D is intersected by an acoustic beam of width W, as is shown in Fig. 5.36, the light diffracted on the acoustic wave has a maximum provided that the Bragg condition is met:

$$\sin \frac{\theta}{2} = \frac{1}{2} \lambda \frac{f}{v}$$

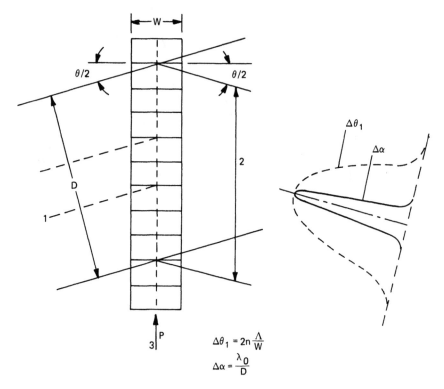

Figure 5.36. Light beam diffraction on acoustic wave: $\theta/2$, angle of light beam incidence; Δd, angular width of diffracted beam, and $\Delta\theta_1$, light beam scanning range. (After M. Barnoski, *Introduction to Integrated Optics*, 1974.)

where $\theta/2$ is the angle of light incidence measured with respect to the acoustic wavefront, $\lambda = \lambda_0/n$ is the wavelength of light falling on a medium with refractive index n, and v and f are sound speed and frequency, respectively.

A collimated light beam may be deflected within angle

$$\Delta\theta = \frac{2n\Lambda}{W}$$

where Λ is sound wavelength under variations of acoustic frequency within

$$\Delta f = 2n \frac{v}{\lambda_0} \frac{\Lambda}{W}$$

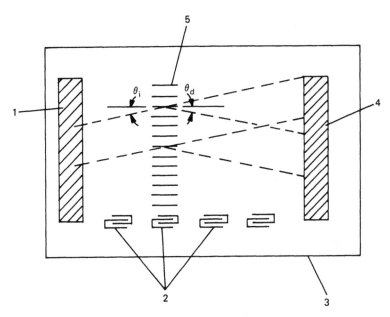

Figure 5.37. Multielement thin-film acoustooptic deflector: (1) light input element, (2) electroacoustic transformers, (3) thin film on piezoelectric wafer, (4) light output element, and (5) acoustic surface waves. (After M. Barnoski, *Introduction to Integrated Optics*, 1974.)

The number of deflector positions is

$$N_x = \tau \, \Delta f$$

where $\tau = D/v$ is the time required for an acoustic wave to cross a laser beam of diameter D.

Crystals of $LiNbO_3$ are popular for wafers because it has good piezoelectric and electrooptic constants. On wafers of $LiNbO_3$, effective countercomb wave transducers may be fabricated for excitation of acoustic surface waves with a central frequency of 1 Gcps and a bandwidth of over 25%. Epitaxial films of ZnO and a number of other materials feature good electrooptic and acoustooptic properties.

Acoustooptic deflectors with resolution over 1000 positions may be built around thin-film waveguides. To this end, several countercomb wave transducers with different central frequencies may be used, as shown in Fig. 5.37.

As may be seen from the above treatment, integrated optics makes feasible laser radiation controllers with outstanding size, switching frequency, and control power as compared with the three-dimensional version. But specific design of, say, a deflector, is dictated by numerous factors, such as laser power, allowable losses for radiation I/O, and requirements for radiation parameters (divergence, coherence, polarization, etc.).

BIBLIOGRAPHY

Balkanski, M., and Lallemand, P. (eds.), *Photonics*, Gauthier-Villars, Paris, 1975.

Baron, R. C., and Piccrilli, D. T., *Digital Logic and Computer Operations*, McGraw-Hill, New York, 1967.

Barrekette, E. S., Stroke, G. W., Nesterikhin, Y. E., and Kock, W. E. (eds.), *Optical Information Processing*, Vol. 2, Plenum Press, New York, 1978.

Basov, N. G, Nobel Prize Lecture, Usp. Phys. Nauk 85, No. 4, 985, 1965.

Brody, T. P., Davis, D. H., and Szepesi, Z. P., *Thin Film Transitor Arrays for Parallel Optical Logic: The Image Negator*. Technical Digest of the IEDM conference in Washington, D.C., December 1975, p. 204.

Collier, R. J., Burkhart, and Lin, L. H., *Optical Holography*, Academic Press, New York, 1971.

Edelstein, L. A., *Picture Logic for "Bacchus", a Fourth Generation Computer*, Computer Journal 6, No. 2, 1963.

Effective Utilization of Optics in Radar System, Proceedings of the Society of Photo Instrumentation Engineers, *128*, 1977.

Enslow, P. H., *Multiprocessors and Parallel Processing*, Wiley Interscience, New York, 1975.

Lavrishchov, V. P., and Svidzinski, K. K., *Problems of construction of optoelectronic systems for information processing*, Microelectronika 2, No. 1, 1973.

Lee, S., *Optical Data Processing: Fundamentals*, Springer-Verlag, Berlin, 1977.

Morozov, V. N., *Digest of Technical Papers*, CLEA (1977), p. 81, IEEE Cat. No. 78 CH 1281-5QEA.

Preston, Jr., K., *Coherent Optical Computers*, McGraw Hill, New York, 1972.

Shaefer, D. H., and Strong, J. P., The Computer, Proc. IEEE 65, No. 4, 129-138, 1977.

Smith, J. H., *Digital Logic*, Newnes-Butterworths, London, 1971.

Special Issue on Optical Computing, Proc. IEEE 65, No. 4, 1977.

Szepesi, Z. P., Brody, T. P., and Davies, D. H., *Basic Circuits for Optical Parallel Processing*, International Conference on Optical Computing in Research and Development, Proc. Visegrad, 4-9 October, 1977, p. 267.

Tamir, T., *Integrated Optics*, Second Edition, Springer-Verlag, New York, 1979.

Taylor, H. T., *Guided Wave Electrooptic Devices for Logic Computation*, Applied Optics 17, No. 10, 1493, May 15, 1978.

Texas Instruments, *The TTL Data Book for Design Engineers*, 1976.

Volodin, E. B., and Svidzinski, K. K., The feasibility of construction of integrally controlled transparencies for an optical coding technique and connection, Autometriya, No. 4, 1977.

Woollons, D. J., *Introduction to Digital Computer Design*, McGraw-Hill, New York, 1972.

6

Optoelectronic Control Operator Processor: Design Philosophy

6.1 BASIC NOTIONS OF NUMBER SYSTEMS

Operations in digital computers are executed over numbers represented as special codes in the number system used in a given computer. A method for the naming and representation of numbers by means of symbols with definite quantitative values is referred to as a *number system*. Symbols representing numbers are called *digits*. Depending on the method of number representation of digits, number systems are classified as positional and nonpositional.

A number system in which the quantitative value of each digit depends on its place or position in the number is referred to as positional. In the positional system any number is represented as a sequence of digits

$$A = \pm a_{m-1} a_{m-2} \cdots a_k \cdots a_1 a_0 . a_{-1} \cdots a_{-\ell}$$

Places enumerated by the indices k (in this case, within the limits $-\ell \leq k \leq m - 1$) are termed the position. The sum $m + \ell$ is equal to the number of positions in a number whose integer and fractional parts have m and ℓ positions, respectively. Each position a_k in the above sequence takes one N possible values, i.e., $N - 1 \geq a_k \geq 0$. The number N of different digits in the positional number system is referred to as the radix N, which defines the name of the system.

The binary system is most convenient for arithmetic and logic operations on numbers in electronic digital computers, owing to the following advantages.

First, binary numbers may be represented as reasonably simple and reliable bistable electronic elements. One state is regarded as 1, another as 0. Evidently, this holds also for optoelectronic processors.

Number system	Addition	Subtraction	Multiplication	Division
Decimal	11.625 +9.125 ⎯⎯⎯ 20.750	11.750 -9.125 ⎯⎯⎯ 2.625	13.50 X5.25 ⎯⎯⎯ 6750 +2700 6750 ⎯⎯⎯ 70.8750	35⎪7 -35⎪5 ⎯⎯ 00⎪
Binary	1011.101 +1001.001 ⎯⎯⎯⎯ 10100.110	1011.110 -1001.001 ⎯⎯⎯⎯ 10.101	1101.10 X101.01 ⎯⎯⎯⎯ 110110 +110110 110110 ⎯⎯⎯⎯ 1000110.1110	100011⎪111 -111 ⎪101 ⎯⎯⎯ 00111 -111 ⎯⎯⎯ 000⎪

Figure 6.1. Examples of arithmetic operations in decimal and binary number systems.

Second, arithmetic and logic operations over binary numbers are rather simple. Indeed, addition and multiplication tables for one-bit numbers are absolutely simple:

Arithmetic addition:

$0 + 0 = 0$

$0 + 1 = 1$

$1 + 0 = 1$

$1 + 1 = 10$

Arithmetic multiplication:

$0 \times 0 = 0$

$0 \times 1 = 0$

$1 \times 0 = 0$

$1 \times 1 = 1$

Multipositional binary numbers are added, subtracted, multiplied, and divided according to the same rules as those used in the decimal system (Fig. 6.1).

Input			Output	
First number	Second number	Carry	Sum	Carry
0	0	0	0	0
0	0	1		
0	1	0	1	0
1	0	0		
1	1	0		
1	0	1	0	1
0	1	1		
1	1	1	1	1

Figure 6.2. Tabular description of a one-bit binary adder.

As arithmetic addition plays a leading role in computations, let us discuss it in more detail. Addition in one position requires three digits: two in the respective positions of both addends and a carry from a lower position. Addition results in two digits: a digit for position in the sum and a carry to the next highest position. Therefore, a one-digit adder has three inputs and two outputs, and its operation may be described by the table in Fig. 6.2.

A circuit implementing this table is shown in Fig. 6.3. It generates sum and carry depending on a combination of inputs, and it is memoryless. A full multidigit adder is constructed of as many one-digit adders as there are addend positions by connecting carry outputs of lower positions with carry inputs of higher positions.

Conventional positional systems (binary, decimal, etc.) are not always convenient for arithmetic operations over digital pictures because carriers under unfavorable conditions propagate from one position to another along a word.

There are number systems having either no carry, or a carry propagating at most along one or two positions. In this case, number representation contains the redundant information necessary to eliminate "long" carries. Such systems may be exemplified, for instance, by the residue, which is based on N mutually disjoint positive integers termed *moduli*, m_1, m_2, ... , m_N. Any number

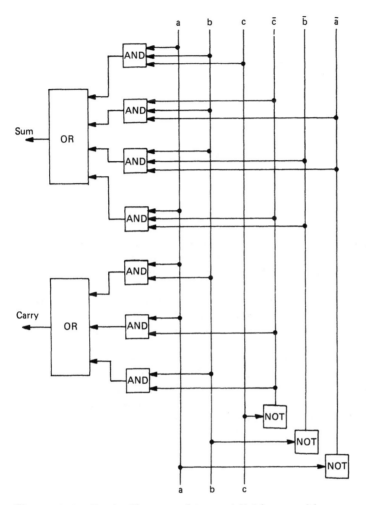

Figure 6.3. Logic diagram of a one-bit binary adder.

is represented by positive integer residues of division by modulus m_i. The division residue is denoted R_{mi}, and N residues $(Rm_1,$ $Rm_2, \ldots, Rm_N)$ of divisions by different moduli represent any integer within the range 0 to M - 1, where $M = \Gamma_{i=1}^{n} m_i$. Numbers greater than M are not defined in this system.

Consider examples of various operations in the residue system with moduli 5, 7, 9, and 4. These numbers are mutually disjoint,

Residue Addition

		5	7	9	4
19	→	4	5	1	3
+87	→	+2	+3	+6	+3
106	←	1	1	7	2

Figure 6.4 Example of addition in a residue system.

their product is M = 1260, and thus the range of representable numbers is 0 to 1259. Figure 6.4 gives an example of the addition of two numbers in the residue system. Number 19 in residues is represented as 4, 5, 1, and 3, which are the residues of the division of 19, respectively, by 5, 7, etc. Number 87 in residues is 2, 3, 6, and 3. Now, each column in Fig. 6.4 is added independently, but the sum is expressed through the residue with respect to the corresponding modulus. For example, 4 + 2 = 6, but the residue of 6 with respect to modulus 5 is 1; similarly, 5 + 3 = 8, but the residue with respect to 7 is 1, etc. Summation gives 1, 1, 7, and 2, which implies that the sum 106 divided by 5 results in residue 1, division by 7 gives 1, etc. The result may be verified by representing the conventional sum in residues.

An example of subtraction is given in Fig. 6.5, where the subtrahend is transformed into residues, and then complements are determined for each residue with respect to the appropriate moduli. The complement of 4 to 5 is 1, that of 1 to 7 is 6, etc. This operation is followed by addition as above.

Multiplication in residues is shown in Fig. 6.6. Both factors are transformed into residues, and the product of residues in each column is expressed through the residue with respect to the corresponding modulus. Residues 1, 2, 3, and 0 represent the decimal product 156.

As may be seen from the above examples, computations in residue classes suit optical digital processors well. The absence of carries permits one to carry out operations in parallel because the results of each basis are not interrelated.

Of interest is, undoubtedly, a positional double redundancy system in which each digit is represented by two symbols (half-positions). Owing to this, the system is referred to as a half-position number system (HPS). An HPS of modulus 2 preserves

Residue Subtraction

		5	7	9	4
106	→	1	1	7	2
-99	4 1 0 3 →	+1	+6	+0	+1
7	←	2	0	7	3

Figure 6.5. Example of subtraction in a residue system.

the polynomial form of a binary number, but each digit consists of two parts, $h_i = \beta_i 2^1 + d_i 2^0$; i.e., each position has two digits; zero = 00, one = 01, two = 10, and three = 11. Normally binary position is written as "single valued," with higher half-position 0, and complementing or inversion inverts only the lower half-position.

In the half-position system, arithmetic operations are also performed without carries. An adder algorithm may be based on logic AND and mod 2 addition if the results of the first operation are left-shifted by one half-position and logically added to the results of the second operation, i.e., if

$$C' = (a \oplus b) \; V \; (a \cdot b)_{\leftarrow \frac{1}{2}}$$

where $(a \oplus b)$ is a position-by-position sum, and $(a \cdot b)_{\leftarrow 1/2}$ is the result of logical multiplication followed by a half-position shift.

Residue Multiplication

		5	7	9	4
13	→	3	6	4	1
X12	→	X2	X5	X3	X0
26					
+13					
156	←	1	2	3	0

Figure 6.6. Example of multiplication in a residue system.

HPS redundancy may result in numbers not written in the normal form. Number 2 may be written (10) instead of (0100); 3 may be written (11) instead of (0101). This prevents summation of more than two pictures, because the resulting combinations (10) and (11) should be reduced to the normal form prior to the next arithmetic operation.

A number system termed *sign-positional representation* (SPR) is very attractive for an optoelectronic digital processor. Sign-positional representations are positional redundant systems in which each position is represented by a signed digit. Any positive integer greater than 2 may be used as a modulus. If in conventional systems with integer modulus (N > 1) each digit may take any of N values (0, 1, 2, ..., N - 1), in the sign-positional representation each digit may take g values according to $N + 2 \leq g \leq 2N - 1$. In SPR, both positive and negative digits are used, and the number of digits used is greater than in conventional positional systems. Thus, SPR has redundancy because each position in a sum or a difference is a function of only two neighboring positions in each number involved. Therefore, the sign-positional representation enables completely parallel additions and subtractions, the execution time always equal to the time of adding or subtracting digits in one position.

As multiplication or division is reducible to a sequence of additions or subtractions and shifts, it might be readily seen that all four basic arithmetic operations may be executed in an optoelectronic processor in parallel.

The above number systems allow one to very effectively exploit the parallel data processing capabilities of optoelectronic processors, the choice of most effective number systems still far from exhausted.

6.2 METHOD OF CONTROL OPERATORS IN OPTOELECTRONIC DIGITAL PROCESSORS

The need for optoelectronic processors may be substantiated as follows. Today's electronic computers have certain fundamental drawbacks because they store and handle data in electronic circuits whose functions are rigidly fixed and cannot be changed. Among such dedicated logical and functional circuits, which are basic computer building blocks, one may list adders, counters, inverters, decoders, etc.

It is, obviously, impossible to design as many units and circuits to enable execution of all or at least a majority of computational operations in one clock cycle, i.e., during the time of access to a corresponding unit or circuit. Therefore, computers make use of a limited set of different basic operations that are hard wired and executed without software control.

In consequence, general-purpose computers intended for a broad class of tasks widely use software control, and since various computational operations are executed as sequences of basic operations, available computer speed and memory area are used ineffectively, and sometimes cannot be used at all. A good example of this situation is that even very fast computers are incapable of real-time pattern recognition.

Dedicated computers use their speed and hardware more effectively. They are designed for specific applications. Some units in special-purpose computers may be more evolved; others are simpler. For example, if a single task is executed with different data, a program may be stored in read-only memory; if computations are reduced to standard simple operations, software control may be simplified or substituted by hard-wired circuitry. Therefore, dedicated computers are much simpler and require less hardware. However, the necessity to use essentially different computers for slightly differing tasks is a serious drawback. The optoelectronic processor relying on control operators to a large measure may be free of the above defects.

Computations are known to consist of a series of transforms over operand sets (input variables). One of the advantages of optical data processing is the possibility of arranging operands in the form of a two-dimensional picture. Mathematically, this is as follows:

$$X_2(\eta, \xi) = \hat{\phi}_1 X_1(\eta, \xi)$$
$$X_3(\eta, \xi) = \hat{\phi}_2 X_2(\eta, \xi)$$
$$\vdots$$
$$X_n(\eta, \xi) = \hat{\phi}_{n-1} X_{n-1}(\eta, \xi)$$
$$Y = X_n(\eta, \xi) \tag{6.1}$$

where X_i are input variables in the ith cycle and output variables in $(i - 1)$st cycle; $\hat{\phi}_i$ is the ith operator corresponding to given basic operations; X and $\hat{\phi}$ represent a two-dimensional luminous picture consisting of light and dark points, ones and zeros; and Y is the result of a computation consisting of n serial basic operations.

Expressions like (6.1) may be implemented through the method of control operators based on the fact that for a certain representation of operands and organization of the optical data processing

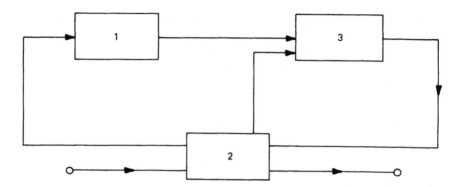

Figure 6.7. Possible version of a control-operator optoelectronic processor: (1) optical memory-storing control operators $\hat{\phi}$, (2) optical buffer main memory, and (3) optical data processing unit built around optically controlled SLM.

path, to each logical or arithmetic operation an operator (or a set of operators) $\hat{\phi}$ is dependent only on operation and is independent of input variables. Consequently, optical memory storing operators in two-dimensional form, a data processing optical path, which is the same for all the basic operations, and a buffer optical memory for exchanges of processed data are required. The difference between the control operator method and that of tabular computations in which memory stores operation results lies in the fact that the former stores types of operations rather than their results. Electronically speaking, operators perform functions of logical and arithmetic electronic devices.

A possible version of an optical processor having such a path is shown in Fig. 6.7. Memory 1 stores operators $\hat{\phi}$, memory 2 is the buffer optical main memory, and unit 3 performs data processing. The processor operates as follows. An array of input variables stored in memory 2 is projected over unit 3 into which operators also enter from memory 1. Using superposition of variable and operator pages, unit 3 performs corresponding data processing. The result is read out of unit 3 and written again into memory 2, out of which it may be fetched for further processing or output to external devices. It should be noted that unit 3 is implemented by means of fast digital optically controlled space-time light modulators (SLM).

It is expedient to implement memory 1 by holographic memories that are sufficiently evolved. Two-dimensional information may

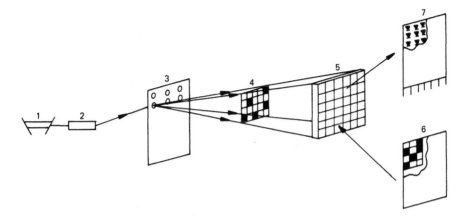

Figure 6.8. Functional diagram of optoelectronic data processing unit: (1) laser, (2) deflector, (3) hologram matrix, (4) restored operator picture, (5) digital optically controlled SLM, (6) digital picture, and (7) input surface of succeeding SLM.

be written on holograms in discrete, analog, or symbolic form. The application of holographic memory for operator storage is also due to its other merits, such as high noise immunity and rapid page reading.

Figure 6.8 shows a functional diagram of an optoelectronic processor. A digital optically controlled SLM is the basic element of the optical discrete data processing circuit implementing the operator. Processing in SLM may be conveniently realized through the following functions:

$$y_{ij} = \begin{cases} 1 & \text{if} \quad z_{ij} = 1 \quad \text{and} \quad x_{ij} = 1 \\ 0 & \text{otherwise} \end{cases} \tag{6.2}$$

$$y_{ij} = \begin{cases} 1 & \text{if} \quad z_{ij} = 0 \quad \text{and} \quad x_{ij} = 1 \\ 0 & \text{otherwise} \end{cases} \tag{6.3}$$

where $\{x_{ij}\}$ is the processed picture, $\{z_{ij}\}$ is the operator, and $\{y_{ij}\}$ are output variables.

Equalities (6.2) and (6.3) mean that if optical properties of reflecting (passing) SLM are modulated by the picture $\{z_{ij}\}$, reflected (passing) light will be only in those SLM cells to which

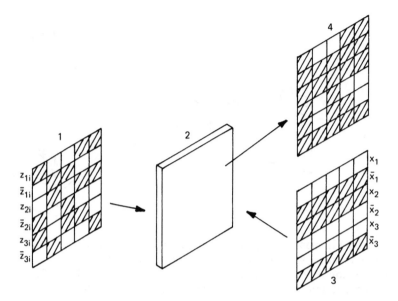

Figure 6.9. Example of transformation of input variable $\{x_i\} = 101$ by operator matrix

$$\{z_{ij}\} = \begin{bmatrix} 0 & 1 & 1 & 0 & 0 \\ 1 & 0 & 1 & 0 & 0 \\ 0 & 1 & 0 & 0 & 1 \end{bmatrix}$$

Variables and operator are in the paraphase code. (1) Input operator field, (2) optically controlled reflecting SLM, (3) input digital field, and (4) result.

light comes from both z_{ij} and x_{ij}, or from x_{ij} only. Any of the definitions of (6.2) and (6.3) may be taken as a function of the operator $\hat{\phi}$, the choice being dictated by the type of SLM used and by the completeness requirement for Boolean functions. Figure 6.9 gives an example of applying transform (6.2) to the variable $\{x_i\} = (101)$ by the matrix

$$\{z_{ij}\} = \begin{bmatrix} 0 & 1 & 1 & 0 & 0 \\ 1 & 0 & 1 & 0 & 0 \\ 0 & 1 & 0 & 0 & 1 \end{bmatrix}$$

Variables and operators are represented in the paraphase code, a light cell designating the presence of light and a dark one

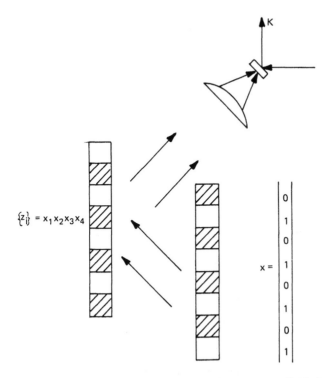

Figure 6.10. Optical scheme and operator field for conjunction $K = x_1 x_2 x_3 x_4$. A cell of inverting SLM is placed in the lens focal plane.

designating its absence. Reflected light is only in those cells on which light falls from both sides.

Consider examples of calculating elementary conjunctions and disjunctions by the control operator method. These operations are fundamental to many logic and arithmetical functions. Let an elementary conjunction $K = x_1 x_2 x_3 x_4$ be desired, where x_i are binary variables and K is the result of conjunction. To compute conjunction, one has to make use of the paraphase representation of binary variables, i.e., to represent 1 as 10 and 0 as 01.

The optical scheme of such an operation and the operator field for computation of conjunction $K = x_1 x_2 x_3 x_4$ are shown in Fig. 6.10. Light is reflected only by those SLM cells that have light from both sides, i.e., from operator and numerical fields. The cylindrical lens collects light in the vertical direction. The presence of light

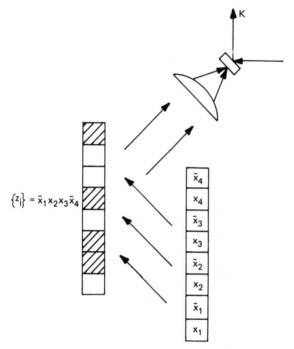

$\{z_i\} = \bar{x}_1 x_2 x_3 \bar{x}_4$

Figure 6.11. Optical scheme and operator field for conjunction $K = \bar{x}_1 x_2 x_3 \bar{x}_4$. A cell of inverting SLM is placed in the lens focal plane.

in the focal plane of the lens means that conjunction is 0, and its absence means that it is 1. Obviously, K = 1 only on the set (1111); otherwise, light is reflected at least by one SLM cell; i.e., there will be light in the lens focal plane, which means that K = 0. In the focal plane there is a detector, a cell of digital controlled SLM, reflecting if there is no light on its input. Stated differently, the cell performs signal inversion: $\bar{K}$ is generated in the focal plane, and the light reflected by the detector cell implements K.

    The order of significant cells in operator and numerical fields is of importance for the execution of logic and arithmetic operations through the control operator method. Thus, if we have assumed for numbers that 1 is represented first by a light and then by a dark cell, the operator field for conjunction is first represented by a dark and then by a light cell.

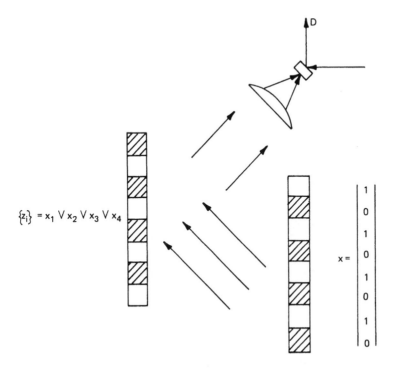

Figure 6.12. Optical scheme and operator field for disjunction
$D = x_1 \lor x_2 \lor x_3 \lor x_4$.

If an elementary conjunction involves terms with negation
(e.g., $K = \bar{x}_1 x_2 x_3 \bar{x}_4$), light and dark cells for terms $\bar{x}_i$ should
be reversed. An operator field for conjunction $K = \bar{x}_1 x_2 x_3 \bar{x}_4$ is
shown in Fig. 6.11. It is clear that there is no reflected light
in the focal plane only over variable set 0110; otherwise, there
is light reflected at least by one cell, i.e., $K = 0$.

Disjunction $D = x_1 \lor x_2 \lor x_3 \lor x_4$ is executed similarly. Its
operator field is shown in Fig. 6.12. It is the order of light and
dark cells that makes this field different from that of the above
examples. There is no reflected light only on set $(0,0,0,0)$, and
$D = 0$. On any other set, light intensity in the lens focal plane
differs from zero, and $D = 1$. If negated variables are involved
in disjunction, the order of light and dark cells in the operator
field should be reversed. Figure 6.13 shows the operator field
for disjunction $D = \bar{x}_1 \lor x_2 \lor \bar{x}_3 \lor \bar{x}_4$. It is obvious that there

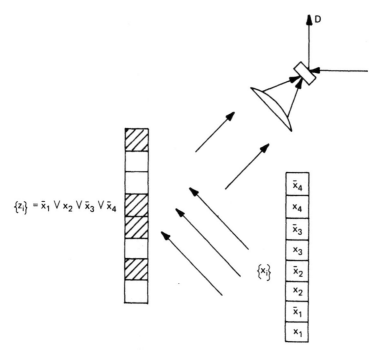

$\{z_i\} = \bar{x}_1 \lor x_2 \lor \bar{x}_3 \lor \bar{x}_4$

Figure 6.13. Optical scheme and operator field for disjunction $D = \bar{x}_1 \lor x_2 \lor \bar{x}_3 \lor \bar{x}_4$.

is no reflected light only on set (1011); otherwise, light falls on the detector resulting in reflected signal, i.e., $D = 1$.

Optical schemes for conjunctions and disjunctions of multiple variables are similar except that, for conjunction inverting, an SLM cell is used as a detector. If for both schemes the same detector is used, which reflects light only if there is light on its input, the schemes in Figs. 6.10 and 6.11 have at their outputs signal $\bar{K}$.

In virtue of the following laws of the Boolean algebra:

$$\overline{\left(\bigwedge_{i=1}^{n} x_i \right)} = \bigvee_{i=1}^{n} \bar{x}_i$$

and

$$\overline{\left(\bigvee_{i=1}^{n} x_i \right)} = \bigwedge_{i=1}^{n} \bar{x}_i$$

one may conclude that any of the schemes shown in Figs. 6.10 through 6.13 may compute both conjunctions and disjunctions of multiple variables. Obviously, digitally controlled N X N SLM computes N various conjunctions or disjunctions of N/2 variables in one cycle. In this case, the schemes of Figs. 6.10 through 6.13 require rows of N detectors.

Logic functions are known to be representable in the disjunctive normal form, or logic sum of products. Therefore, unit 3 processing operands may consist of only two interrelated SLM forming a common data processing path. In doing so, the first optically controlled SLM may be used as an operating plate, implementing, depending on the operator, a set of elementary conjunctions, such as

$$f_1 = x_1 x_2 \cdots x_i \cdots x_{n-1} x_n$$

$$f_2 = x_1 \bar{x}_2 \cdots \bar{x}_i \cdots x_{n-1} \bar{x}_n$$

$$\vdots$$

$$f_k = \bar{x}_1 x_2 \cdots x_i \cdots \bar{x}_{n-1} x_n$$

Another SLM may be used as a switching plate implementing a set of disjunctions, such as

$$y_1 = f_1 \vee f_2 \vee \cdots \vee f_i \vee \cdots \vee f_{n-1} \vee f_n$$

$$y_2 = \bar{f}_1 \vee f_2 \vee \cdots \vee \bar{f}_i \vee \cdots \vee f_{n-1} \vee f_n$$

$$\vdots$$

$$y_k = f_1 \vee \bar{f}_2 \vee \cdots \vee f_i \vee \cdots \vee \bar{f}_{n-1} \vee \bar{f}_n$$

In some particular cases, three or more SLM may turn out to be expedient. The third SLM may, for example, distribute the result between certain cells of the two-dimensional discrete page in order to make further processing more convenient.

Thus, at each clock cycle three SLM plates enable implementation of practically any logical function of up to 10^4 binary variables, all the logical relations defined by two-dimensional operators stored in the holographic memory.

6.3 FEATURES OF DATA PROCESSING IN OPTOELECTRONIC CONTROL-OPERATOR PROCESSORS

Computer functional capabilities are based on two fundamental ideas. The first is that the program should be input through

the same external devices and stored in the same memory as
the source data. This enables rapid computer readjustment
from one task to another without changes in computer structure and interconnections, thus making it a universal computing
instrument.

Another important idea is that program instructions are coded
as numbers that formally do not differ from the numbers handled
by the computer.

This allows one, in the course of executing one instruction, to
regard another (or the same one, in a particular case) as a number, to send it to the arithmetic unit, to handle it, and to return
to its memory location in a changed form. When the control unit
accesses this location the next time for instruction rather than
for operand, it executes the modified instruction rather than the
original one. Thus, the execution of a program may be accompanied by its modification.

The availability of control transfers on the list of control unit
operations is almost as important as the above feature. If all the
instructions of a program were to be executed in a sequence according to their position in the memory, each instruction would
be executed only once. In doing so, the computer would be able
to deal with comparatively simple tasks whose execution requires
no more operations than the amount of instructions that may be
simultaneously stored by the computer memory. However, it would
be unreasonable to execute these tasks on a computer because,
evidently, the generation of each computer instruction and writing
on punched tape or a card, or any other carrier, requires at least
as much time as manual execution of this single operation.

It is quite obvious that the optical processor will be also able
of doing this. Indeed, input/output (I/O) and data exchange
are done in the same manner as in the electronic computer, and
control transfers may be always added to the operation list of
the optical processor control unit without any constraint. What
is more, the optical processor has some important features supporting high throughput for most diverse tasks.

Data are processed by digital optically controlled SLM whose
capacity in digital processing may be 10^4 bits and more, with
processing executed in parallel. Such a high degree of parallelism allows one to employ matrix and vector computations. Recent
attempts to reproduce electronically this intrinsically optical feature should be noted. High-throughput computers based on vector data processing have been and are designed (Cray-1, Star-
100, and ASC), but the execution of inherently sequential tasks
results in a dramatic throughput reduction.

The control operator method also allows one to parallelize significantly an instruction execution. If in powerful computers the amount of hard-wired logic and other operations may be several hundreds (as a rule, 100 to 200), in the optical processor the number of control operators may run to 10^4. Indeed, since the capacity of available holographic read-only memories is estimated as 10^8 bits and the size of the SLM page is about 10^4 bits, the number of basic operators may be about 10^4 bits. This brings about two important features: first, the generation of a task execution program based on such a set of operators may lead to a shorter algorithm; second, there is a principle possibility of effective execution of both matrix and sequential tasks. It should be noted that parallelization with respect to instructions of up to 10^4 optical processor operators is equivalent in the electronic computer to the availability of the same number of different logic and other circuits and units. This is impractical because it leads to a sharp increase in complexity and size, handicaps interconnections between individual electronic devices, etc. Some operators may be oriented to the most frequent tasks.

One of the major distinctions of the optical processor from the electronic computer is the absence of dedicated units and circuits. Thus, the necessity to implement interconnections between individual logic, arithmetic, etc., devices disappears. The common data processing path in SLM substitutes various dedicated units and circuits. If operator memory is implemented by a reversible medium, in the course of computations new basic operators may be generated and existing ones may be modified with an allowance for current results. Consequently, in such a processor, operators may be dependent on processed variables. The implementation of operand-dependent basic operators has no analog in traditional electronic computers where this is equivalent to online addition of new computing facilities or rearrangement of existing ones. Such an implementation enables simple rearrangement of processor operation at the logic level and makes it very flexible and effective. The rearrangeability of the optical processor at both the program and logic levels and peculiarities of optical processing of two-dimensional data arrays, possibly, will sometimes enable the application of special computational techniques to very involved and frequently occurring functions that present serious difficulties to conventional electronic computers. For example, associative data processing might permit one to perform complicated transforms with little hardware.

The optical processing path with control operators consists of several interconnected SLM, of which one computes conjunctions,

another computes disjunctions, and the third distributes data between appropriate main memory locations. SLM perform sequential
data processing, i.e., the results of the first SLM are sent into
the second one, etc. All the SLM make up one path; therefore,
the importance of coupling them optically is evident.

Cylindrical lenses are the simplest means to effect such coupling.
They collect light from rows or columns and implement disjunction
or conjunction of processed variables as required by the control
operators. To make the Boolean basis complete, one should add
inversion, which is done by a row of inverting SLM whose electronic logic operates so that the controlled surface does not reflect light if light falls on a cell of the controlling surface. For
further processing, light collected in the focal plane after SLM
should be again distributed along a row or column. This operation is also done by a cylindrical lens. Optical signals are amplified at reflection at the expense of radiation falling on the controlled SLM surface. This method of optical SLM coupling is simplest for implementation, but has the serious drawback of collec-
ting light from the entire row or column in all cases even if a
simple operation is required, say, conjunction of two variables.
In this case only four cells of a 100-bit row are used, and consequently, the capabilities of parallel processing deteriorate.
For example, it is impossible to compute conjunctions of a variable number of elements. Essential gain may be obtained by providing electrical coupling of neighboring cells of a row or column
in digital optically controlled SLM, which has some distinctions as
compared with conventional types of SLM.

First, if control light falls on an SLM cell, its optical characteristics change, as well as those of other column or row cells.

Second, the number of cells so coupled may vary depending
on an optical or electrical signal, called the *setting signal*. In
this case, an optical signal arriving at one cell may control the
length of a row or column. Technologically, these SLM do not
differ from the latrix: the individual integrated circuits (IC) of
each cell should be interconnected by electronic keys whose state
(on or off) depends only on setting signals. Such devices for
optical SLM coupling are equivalent to a matrix of cylindrical
lenses with a variable number of elements in rows.

Figure 6.14 shows schematically such SLM consisting of (1) a
controlling surface, (2) a controllable latrix surface, (3) microelectronic cells controlling the modulating layer, and (4) electronic keys connecting neighboring cells. This SLM operates
like the conventional digital one, but a signal on the controlling

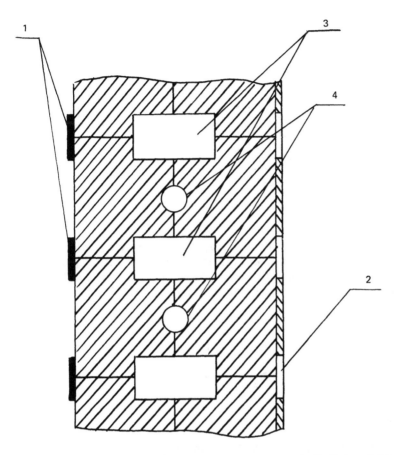

Figure 6.14. Scheme of an optically controlled digital neighbor-hood SLM: (1) control surface photodiodes, (2) output surface-controlled light gates, (3) microelectronic cells controlling re-flectance of light gates, and (4) electronic keys connecting neighboring cells.

surface of a cell switches both it and its neighbor if the key be-tween them was not set "on" previously.

Thus, one SLM and the above SLM with interaction between neighbors in row and column may simultaneously compute the following set of conjunctions or disjunctions:

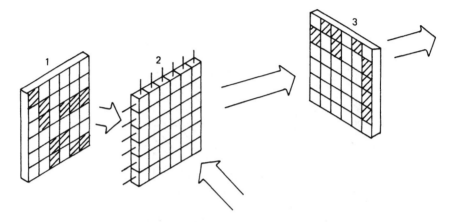

Figure 6.15. Lensless data processing path relying on an intermediate neighborhood SLM: (1) controlled SLM output picture, (2) neighborhood SLM with rows or columns set to execution of conjunctions K_i or disjunctions D_i, and (3) resulting picture.

$$K_1 = a_1 a_2 \qquad\qquad D_1 = a_1 \vee \bar{a}_2 \vee a_3$$

$$K_2 = a_3 \bar{a}_4 \bar{a}_5 \qquad\qquad D_2 = a_4 \vee a_5$$

$$K_n = \bar{a}_i \dots a_{Nc} \qquad D_n = \bar{a}_i \vee \dots \vee \bar{a}_{Nc}$$

by means of a single row and column; $Nc = N/2$, where N is row or column size.

It goes without saying that the number of possible conjunctions and disjunctions defined by the setting signal is very high. Setting signals should be sent by the control unit and stored in the memory. They may be, obviously, changed when desired rather than at each computational cycle.

A possible block diagram of a data processing unit is shown in Fig. 6.15. Light modulated by a control operator and numerical pictures goes from SLM (1) to a strip digital SLM with interaction between neighbors (2) to which setting signals were sent. The SLM may be set, for example, by means of a matrix of address buses. With matrix addressing, each column or row is set in its turn to perform conjunctions (K_i) or disjunctions (D_i).

Read-out data are projected on the controlled surface of SLM 3 for further processing. When used as a coupling and processing unit, a digital SLM with controllable connections in rows and columns enables two-dimensional processing of numbers represented in a rowwise form.

6.4 TYPES OF CONTROL OPERATORS FOR BASIC LOGIC AND ARITHMETIC FUNCTIONS

Electronic computers execute a series of logic operations and instructions, of which the most popular are logic addition, logic multiplication, comparison, inversion, left or right shift of one position, and mod 2 addition. Some of them are executed by the same or slightly augmented circuits (e.g., mod 2 addition and inversion), and others are executed by separate electronic circuits (e.g., shift and logic addition, logic addition, and logic multiplication). In the optoelectronic processor all the operations are realized in one clock cycle (i.e., in a single access to the operator memory) within the same optical path and with the same elements, different operations requiring different operators to be projected. Consider how the optoelectronic processor executes some of these operations.

6.4.1 Mod 2 Addition

For the sake of clarity, consider mod 2 addition of two 3-bit numbers, $A = a_1a_2a_3 = 101$ and $B = b_1b_2b_3 = 110$, and denote by $Q = q_1q_2q_3$ the result, which is computed through the expression $q_i = \bar{a}_ib_i \vee a_i\bar{b}_i$, and, in our example, is $Q = q_1q_2q_3 = 011$. Figure 6.16 shows a circuit implementing mod 2 addition. The operator $\hat{\phi} = \{z_{ij}\}$ corresponding to mod 2 addition is projected on the controlling surface of the optical digital SLM. It changes reflecting properties of the controlled surface onto which added numbers are projected. Each column implements $\bar{a}_ib_i$, and the jth column gives no light only if input signals do not coincide with any light cell in the column. Signals reflected by all the elements of each column are collected by a cylindrical lens. After the inverter, there will be light only if the column reflects no signal. If inverters are combined in pairs, one position corresponds to each pair. Hence, light in any of the inverters means that, in this position, $q_i = 0$; otherwise, $q_i = 1$.

Figure 6.17 shows the position of operator fields implementing bit-by-bit addition of two numbers in the general structure of a

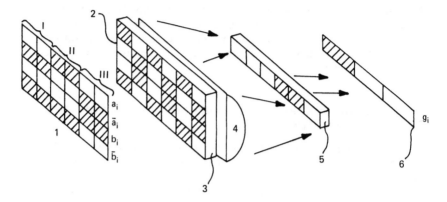

Figure 6.16. Optical scheme and operator fields for mod 2 addition of two 3-bit numbers: (1) input variables in the paraphase code, (2) operator field for mod 2 addition, (3) digital optically controlled SLM, (4) cylindrical lens, (5) row of inverters, and (6) result.

processor consisting of two optically connected SLM. Mod 2 addition of three-, four-, n-bit numbers is done in much the same way. Control operator pictures for bit-by-bit addition of three numbers may be seen in Fig. 6.18.

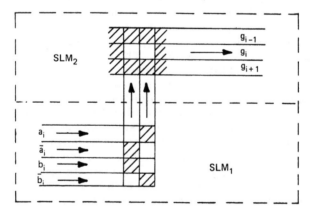

Figure 6.17. Operator fields for bitwise addition of two numbers in an optoelectronic path consisting of two optically coupled digital SLM. Operator fields are shown for both SLM.

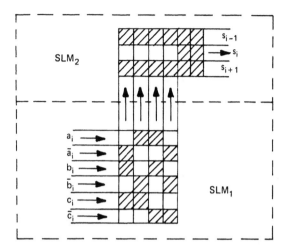

Figure 6.18. Operator fields for bitwise addition of three numbers in an optoelectronic path consisting of two optically coupled digital transparencies. Operator fields are shown for both transparencies.

6.4.2 Shift on Given Number of Positions

An optoelectronic processor may perform right and left shift on any desired number of positions in a single clock cycle. For the sake of example let us see how a three-bit number $A = a_1 a_2 a_3$ is right shifted in one position. The shift is accomplished by two optical digital SLM, SLM-1 and SLM-2, onto which operator fields are projected as shown in Fig. 6.19. Input variables are projected onto the first SLM in the form of horizontal light stripes. Light is reflected only by those cells in which the light of input variables coincides with the light of the operator. Light reflected by the cell is expanded by a cylindrical lens into a column, which is projected on the following SLM in the form of vertical stripes. The operator projected on this SLM modulates the columns, and the output light reflected by the second SLM carries the number $B = b_1 b_2 b_3 b_4$, which is the number $A = a_1 a_2 a_3$ right shifted in one position.

The above and Fig. 6.19 demonstrate that one may shift numbers in any number of positions by appropriate selection of the operator field projected on the second SLM.

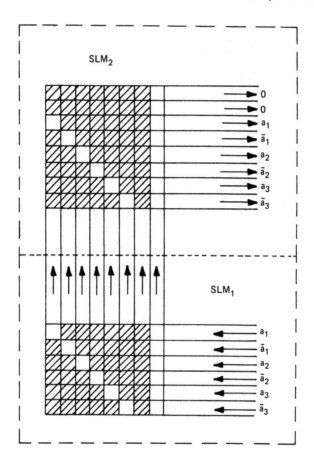

Figure 6.19. Operator fields for a one-bit right shift.

6.4.3 Number Inversion

Number inversion is done by means of operator fields, as shown in Fig. 6.20.

6.4.4 Decoding

A circuit that assigns a signal on a separate output to each input set is referred to as a *decoder*. Generally, a decoder has n inputs

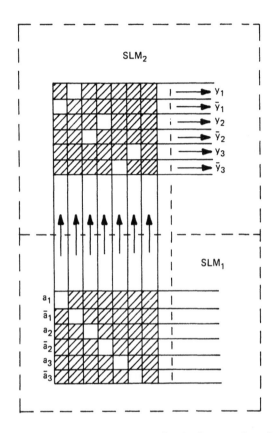

Figure 6.20. Operator fields for number inversion.

and 2^n outputs and transforms a coded word into a control signal at one of its outputs.

In computers, the decoder is employed for sending signals to control circuits, depending on the input word. For instance, it may transform an operation code into a signal controlling the arithmetic unit. Consider the case of $n = 3$. Decoder operation is described by the following equations:

$$y_1 = \bar{a}_1\bar{a}_2\bar{a}_3 \quad y_3 = \bar{a}_1a_2\bar{a}_3 \quad y_5 = a_1\bar{a}_2\bar{a}_3 \quad y_7 = a_1a_2\bar{a}_3$$

$$y_2 = \bar{a}_1\bar{a}_2a_3 \quad y_4 = \bar{a}_1a_2a_3 \quad y_6 = a_1\bar{a}_2a_3 \quad y_8 = a_1a_2a_3$$

and operator fields for three-input decoding are shown in Fig. 6.21.

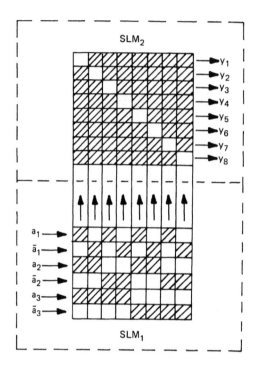

Figure 6.21. Operator fields for three-input decoding.

6.4.5. Computation of Logic Functions

In the course of computations one must implement logic functions of many variables. Since in electronic computers physical elements are assigned to each logic operation, any complicated logic function is hard wired, i.e., is implemented by a set of logic elements. In optoelectronic processors, new operators may be generated as noted above. For the sake of example, Fig. 6.22 shows operator fields implementing the following logic functions:

$$y_1 = x_1 \bar{x}_2 x_3 \bar{x}_4 \vee x_1 \bar{x}_2 x_3 \vee x_2 x_3 \bar{x}_4 \vee \bar{x}_1 x_3$$

$$y_2 = \bar{x}_1 x_2 \vee x_1 \bar{x}_2 \bar{x}_3 \vee x_4 \vee x_1 \bar{x}_4$$

It should be readily apparent from the above examples that a majority of logic operations is computed in the optical processor in one cycle. The number of such operations is practically unlimited and depends only on the capacity of operator memory

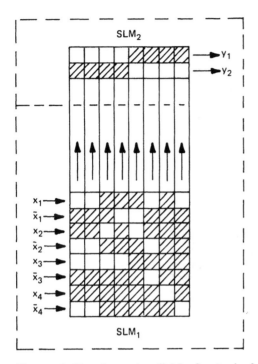

Figure 6.22. Operator fields for logic functions.

within the time interval when stored operator fields are not changed.

6.4.6 Arithmetic Addition

All arithmetic operations rely on some logic functions. The most important arithmetic operation is addition, because in the final analysis, subtraction, multiplication, and division may be reduced to it.

Electronic engineering has a great many methods for arithmetic addition and corresponding methods for adder design. There are parallel, serial, serial-parallel, superparallel, and other adders. Each of them is implemented as a hard-wired electronic circuit that cannot be modified during the computer's lifetime.

In the optical processor, however, the particular form of logical operation depends only on the form of the control operator. The processor, therefore, may perform operations, say, arithmetic

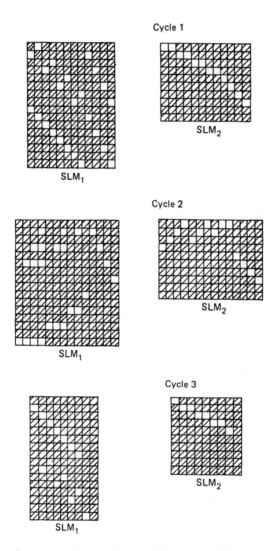

Figure 6.23. Operator fields for the addition of two 4-bit numbers in three clock cycles.

addition, by several techniques. Selection of a particular technique is dictated by the specificity of a problem under consideration. Three techniques of arithmetic addition are presented below as illustrations.

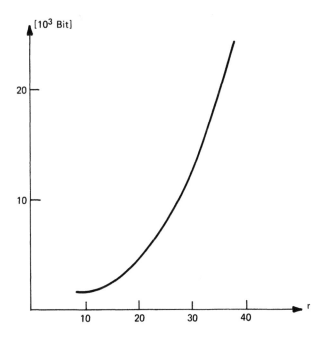

Figure 6.24. Operator memory capacity versus the number of bits for the addition of two numbers in three clock cycles.

Consider first a very simple method for parallel addition of two numbers, $A = a_1 \ldots a_i \ldots a_n$ and $B = b_1 \ldots b_i \ldots b_n$, which may be executed in three cycles. In the first cycle sums and carries are computed for positions

$$q_i = \bar{a}_i b_i \lor a_i \bar{b}_i$$

$$P_i = a_i b_i$$

where q_i is the sum in the ith position, P_i is a carry from the ith to the (i - 1)st position. In the second cycle sets of actual carries are computed

$$\psi_i = P_{i+1} \lor P_{i+2} q_{i+1} \lor \ldots \lor P_0 q_{i+1} q_{i+2} \ldots q_n$$

and the resulting sum $S = S_1 S_2 S_3 \ldots S_n$ is defined in the third cycle:

$$S_i = q_i \bar{\psi}_i \lor \bar{q}_i \psi_i$$

Figure 6.23 represents operator fields of first, second, and third cycles for addition of two 4-bit words through this method. In Fig. 6.24 operator memory capacity is plotted versus the number of

+		111	001	101	-a	+		111	111	111	-a
		010	111	101	-b			000	000	001	-b
I		101	110	000	-q	I		111	111	110	-q
		100	011	010	-p			000	000	010	-p
II		100	111	010	-ϕ	II		000	000	110	-ϕ
	1	1	0	0	-ψ		0	0	1	0	-ψ
	0	0	0	0	-ξ		0	1	1	0	-ξ
III		101	110	000	-q	III		111	111	110	-q
		100	111	010	-ϕ			000	000	110	-ϕ
		001	001	010	-S'			111	111	000	-S'
	1	1	0	0	-α		1	1	1	0	-α
IV	1	010	001	010	-S	IV	1	000	000	000	-S

Figure 6.25. An example of arithmetic addition of two numbers in four cycles.

positions of two numbers added in three cycles. It may be seen that addition of 20-bit numbers requires 2×10^3 memory bits.

The operator memory capacity required for addition may be reduced by means of various methods. Addends may be grouped by m bits, and addition may be performed in four clock cycles, i.e., in four accesses to the operator memory. It may be easily seen that this technique resembles the serial-parallel one, with the difference that carries from groups are considered in one cycle. In the first cycle sums q_i and carries P_i in positions are determined; in the second cycle actual carries within groups ϕ, carries from one group to another ψ, and the ξ attributes of group carries are determined. In the third cycle sums in positions with regard to actual carries S' and actual carries from group to group α are determined, and the final sum S is determined in the fourth cycle.

For the sake of illustration, Fig. 6.25 presents two examples of four-cycle additions for m = 3. In Fig. 6.26, V_i is plotted versus m, where V_i is the operation field memory capacity at the ith cycle. It is clear that the optimal m lies between 3 and 4, and that $V_i \lesssim 10^3$ bits.

The third method relies on multiple one-bit adders; and although it performs addition in n cycles (n = the number of bits

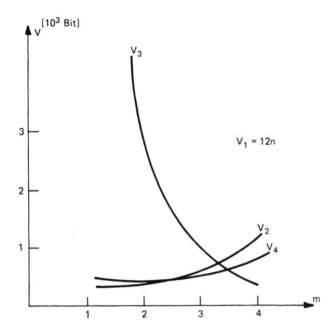

Figure 6.26. Operator field memory capacity versus the amount of numbers in group m for four-cycle addition.

in the word), the large capacity of digital optically controlled SLM enables the concurrent addition of multiple number pairs. Indeed, the following equalities should be realized by a one-bit adder:

$$S_i = \bar{a}_i \bar{b}_i \phi_i \vee \bar{a}_i b_i \bar{\phi}_i \vee a_i \bar{b}_i \bar{\phi}_i \vee a_i b_i \phi_i$$

$$\phi_{i-1} = a_i b_i \vee a_i \phi_i \vee b_i \phi_i$$

where ϕ_i is a carry from the $(i + 1)$st position into ith one, and S_i is the ith position in the sum of A and B. With this method, the operator field is i-independent, and with allowance for the para-phase coding the required operator, memory capacity is $V \sim 40$ bits. Consequently, one N X N SLM may implement $K = (N \times N)/V$ one-bit adders. K number pairs are summed in n cycles, which is equivalent to the simultaneous addition of $M = K/n$ numbers. For $N = 128$ and $n = 32$, $M \sim 10$; i.e., this method is equivalent to the simultaneous addition of ten 32-bit numbers.

Evidently, the gist of this method is parallel processing, which is understood as identical processing of large numerical arrays. However, if the addition of one pair is required, this method is inexpedient.

The selection of this or that method for addition is dictated by the particular class of problems under consideration, but it should again be emphasized that various arithmetic addition operations require only different control operators, with optical processor structure always fixed.

Arithmetic subtraction may be conveniently implemented by means of an inverted code. In doing so, it is not necessary to transform subtrahend B into inverted code, which may be taken into consideration at the first step of subtraction in the control operator field, i.e.,

$$q_i = a_i \bar{\bar{b}}_i \; V \; \bar{a}_i \bar{b}_i = a_i b_i \; V \; \bar{a}_i \bar{b}_i$$

Multiplication may be done through well-known methods of enhanced multiplication. The optical processor may divide two numbers by using bit-by-bit shift and subtraction only.

The above examples illustrate the feasibility of implementing diverse logic and arithmetic operations in an optoelectronic processor by means of the control operator technique. As was noted above, the number of different operations is constrained only by the capacity of holographic operator memory and may be about 10^4 for a memory of 10^8 bits. If, for some operations, operator fields smaller than the optically controlled SLM page are required, several operations, obviously, may be executed at the same time.

BIBLIOGRAPHY

Avizienis, A., *Signed-Digit Number Representation for Fast Parallel Arithmetics*, IRE Trans. Electronic Computers, *10*, No. 3, 1961.

Avizienis, A., *Binary-Compatible Sign-Digit Arithmetics*, AFIPS Conference Proceedings, San Francisco, vol. 26, part 1, 1964.

Basov, N. G., Popov, U. M., Morozov, V. N., et al., *Principles of Construction of Optical Processors with Variable Operators*, Kvantovaya Electronika 3, p. 526, 1978.

Basov, N. G., Popov, U. M., Morozov, V. N., et al., *Methods of Realization of the Optical Processor with Variable Operators*, Kvantovaya Electronika 3, p. 533, 1978.

Chen, D., and Zook, I. D., An Overview of Optical Data Storage Technology, Proc. IEEE *63*, No. 8, 1207, August 1975.

Egorova, L. V., and Rachmanov, V. F., Image Element Arrangement for Parallel Processing Information, in *Optical and Electrooptical Processing of Information*, Moscow, Nauka, 1975.

Hawkins, J. K., and Munsey, C. J., *A Parallel Computer Organization and Mechanization*, IEEE Transactions on Electronic Computer *12*, No. 3, 1963.

Hewlett Packard Components, *Optoelectronics Designers Catalog 1977*, Palo Alto, California, 1977.

Huang, A., *The Implementation of a Residue Arithmetic Unit Via Optical and Other Physical Phenomena*, in Proceedings of the International Optical Computing Conference, April 1975, Washington, D.C., IEEE Cat. No. 75, CH0941-5C.

Huang, A., Tsunoda, Y., Goodman, J. W., and Ishihara, S., *Some Optical Methods for Performing Residue Arithmetic Operations*, in Proceedings of 1978 International Optical Computing Conference, Washington, D.C., IEEE Cat. No. 78, C.H. 1305-2C.

Huang, A., Tsunoda, Y., Goodman, I. W., and Ishihara, S., *Optical Computation Using Residue Arithmetics*, Applied Optics *18*, 149, January 15, 1979.

Johnson, E. O., *Physical Limitations on Frequency and Power Parameters of Transistors*, IEEE International Conference Record, March 1965.

Little, A. D., *World Electro-Optics Industry 1975-1980*, E. O. Components Update, May 1976.

Mnatsakanyan, E. A., Morozov, V. N., Popov, H. M., et al., *Code Processing of Information on Optoelectric Mechanisms*, Kvantovaya Electronika *6*, p. 1125, 1979.

Psaltis, D., and Cassasent, D., Optical Residue Arithmetics: A Correlation Approach, *Applied Optics 18*, 163, January 15, 1979.

Ruggiero, I. F., and Cowell, D. A., *An Auxiliary Processing System for Array Calculations*, IBM Systems Journal *8*, No. 2, 1969.

Semiconductor Group, *Optoelectronics Data Book for Design Engineers*, Fourth edition, Texas Instruments Incorporated, Dallas, 1976.

Szabo, N. S., and Tanaka, R. I., *Residue Arithmetic and Its Application to Computer Technology*, McGraw-Hill, New York, 1967.

Tai, A., Cindrich, L., Ficnup, J. R., and Aleksoff, C. C., *Optical Residue Arithmetic Computer with Programmable Computation Modules*, Applied Optics *18*, No. 16, 2812, August 15, 1979.

7

Examples of Effective Applications of Optoelectronic Processors

7.1 SELECTION OF A MINIMAL OR MAXIMAL NUMBER FROM AN ARRAY OF BINARY NUMBERS

In the preceding chapters, consideration was given to the philosophy of optical processors, examples of logic and arithmetic operations were presented, and features of optical data processing were discussed. It was stressed that an optical processor with control operators allows one to deal with a wide range of problems using a fixed processor structure.

It is of interest to compare the throughputs of the optical processor and the electronic computer. It might be well to point out that two computers may be compared only if they deal with the same scope of problems. Obviously, the narrower the scope, the higher is the throughput attained in it. The version of optical processor discussed is general purpose, and it is right to compare it with general-purpose electronic computers.

An attempt to estimate optical processor effectiveness is taken up on the basis of some specific problems that often occur in computations. Evidently, these tentative estimates are a stimulus for further studies on effective algorithms for the execution of diverse tasks on the optical processor.

Let us consider the problem of searching a minimal number from an array of binary numbers. The algorithm for a maximal number is of a similar nature.

A fast search for a minimal number in a binary array is important in real-time tasks, such as process optimization, control, and planning.

The algorithm proposed is based on parallel bit-by-bit comparison of all the numbers, i.e., on parallel comparison of the same position in all array numbers. Such a comparison, obviously, should be initiated from the first (highest) position.

Let an array of m n-bit numbers be given. Denote by a_{ij} an element of this array where $i = 1, \ldots, m$ and $j = 1, \ldots, n$; indices i and j designating element number and position in it, respectively. To select the minimal number of a set of m numbers, an additional attribute Q_i should be generated for all of them. Then, the processed picture is as follows: General expression:

$$
\begin{bmatrix}
a_{11} & \cdots & a_{1j} & \cdots & a_{1n} Q_1{}^0 \\
\cdots & \cdots & \cdots & \cdots & \cdots \\
a_{j1} & \cdots & a_{ij} & \cdots & a_{in} Q_i{}^0 \\
\cdots & \cdots & \cdots & \cdots & \cdots \\
a_{m1} & \cdots & a_{mj} & \cdots & a_{mn} Q_m{}^0
\end{bmatrix}
$$

Example: $Q_i{}^0$
                              $\downarrow$

$$
\begin{bmatrix}
1 & 1 & 0 & 1 & 1 & 1 & 0 \\
0 & 1 & 1 & 0 & 1 & 1 & 0 \\
0 & 1 & 0 & 1 & 0 & 1 & 0 \\
1 & 0 & 1 & 0 & 0 & 1 & 0 \\
1 & 0 & 0 & 1 & 1 & 0 & 0
\end{bmatrix}
$$

(For the sake of clarity, the general expression is accompanied by an example.) At the beginning of the search all the attributes $Q_i{}^0 = 0$, as shown in the example. The search begins from the highest bit, i.e., from the first column. If all the column a_{i1} consists of 1, this means that all the higher bits are identical (equal to 1) and that the next position should be analyzed. To determine this, let us consider attribute α_1:

$$
\alpha_1 = \bigwedge_{i=1}^{m} a_{i1}
$$

If $\alpha_1 = 1$, all numbers a_{i1} are 1, and one should proceed to the next column, i.e., column a_{i2}, leaving attribute column $Q_i{}^0$ unchanged. If $\alpha_1 = 0$, this means that highest bits of one or more of m numbers are 0 and that the minimal number is among them. At $\alpha_1 = 0$, attributes $Q_i{}^1$ should be computed:

$$Q_i{}^1 = Q_i{}^0 \vee a_{i1}$$

and written instead of $Q_i{}^0 = 0$. As $Q_i{}^0 = 0$, $Q_i{}^1 = a_{i1}$. Computation of $Q_i{}^1$ is followed by elimination in array a_{ij} of the rows, where $Q_i{}^1 = 1$ because the minimal number is among those rows where $a_{i1} = 0$, i.e., $Q_i{}^1 = 0$. General expression:

$$\begin{bmatrix} a_{11} & \cdots & a_{1j} & \cdots & a_{1n} Q_1^0 \ V \ a_{11} \\ \cdots\cdots\cdots\cdots\cdots\cdots\cdots\cdots\cdots \\ a_{i1} & \cdots & a_{ij} & \cdots & a_{in} Q_i^0 \ V \ a_{i1} \\ \cdots\cdots\cdots\cdots\cdots\cdots\cdots\cdots\cdots \\ a_{m1} & \cdots & a_{mj} & \cdots & a_{mn} Q_m^0 \ V \ a_{m1} \end{bmatrix}$$

Example ($\alpha_1 = 1 \wedge 0 \wedge 0 \wedge 1 \wedge 1$):

$$\begin{array}{c} \qquad\qquad\qquad Q_i^1 \\ \qquad\qquad\qquad \downarrow \\ \begin{bmatrix} 1 & 1 & 0 & 1 & 1 & 1 & 1 \\ 0 & 1 & 1 & 0 & 1 & 1 & 0 \\ 0 & 1 & 0 & 1 & 0 & 1 & 0 \\ 1 & 0 & 1 & 0 & 0 & 1 & 1 \\ 1 & 0 & 0 & 1 & 1 & 0 & 1 \end{bmatrix} \end{array}$$

The second column is analyzed in a similar manner. First, $\alpha_2 = \wedge_{i=1}^{m_j} a_{i2}$ is computed, where m_j is the amount of numbers left after the first cycle. If $\alpha_2 = 1$, all the bits are $a_{i2} = 1$ and the next position should be considered. If $\alpha_2 = 0$, attributes Q_i^2 should be computed

$$Q_i^2 = Q_i^1 \ V \ a_{i2}$$

Computation of Q_i^2 should be followed by elimination of rows with $Q_i^2 = 1$, because the minimal number is among rows with $Q_i^2 = 0$.

Computations of α_j and Q_i^j ($i = 1, \ldots, n$) done over all the positions result in a set of 1s and 0s in the attribute column, zeros indicating rows containing the minimal number.

In the numerical example, computation of α_2 for the second position gives $\alpha_2 = 1 \wedge 1 = 1$; therefore, a third position should be considered for which $\alpha_3 = 1 \wedge 0 = 0$. By computing $Q_i^3 = Q_i^1 \ V \ a_{i3}$ and deleting rows with $Q_i^3 = 1$ we obtain

$$\begin{bmatrix} \cdots\cdots\cdots & 1 \\ \cdots\cdots\cdots & 1 \\ 0 & 1 & 0 & 1 & 0 & 1 & 0 \\ \cdots\cdots\cdots & 1 \\ \cdots\cdots\cdots & 1 \end{bmatrix}$$

Obviously, further computations of α_j and Q_i^j do not change Q_i^j; hence, the minimal number is in the third row.

Digital N X N space-time light modulators (SLM) may accommodate $M = N^2/r$ numbers, where r is number length. Therefore,
the maximal number of clock cycles required by the optoelectronic
processor to determine the minimal number is equal to r. Electronic computers make use of a serial comparison of all the numbers; therefore, the search for the minimum of M numbers requires $M/(r_0/r)$ clock cycles, where r_0 is the length of the computer word.

Assuming that clock cycles of a optoelectronic processor and
an electronic computer are equal, determine that the gain is $N^2/r_0 r$. For N = 128 and $r = r_0 = 32$, the gain is 16. If numbers are
shorter than the computer word, the gain is higher; for $r_0 = 32$
and r = 16 or 8, it is, respectively, 32 and 64.

7.2 MINIMIZATION OF LOGIC EXPRESSIONS

The design of any computer unit intended for the execution of a
group of logic and arithmetic operations pays special attention to
hardware reduction. As all the logic and arithmetic operations
rely on the Boolean algebra, minimization of appropriate logic expressions plays a prominent part in the solution to this problem.

Any logic function may be represented by various equivalent
logic formulas. The diversity of formulas expressing a logic function defines the diversity of their representations in terms of various logic operations.

The so-called normal forms of complex logic functions are convenient for practical purposes. Their construction is based on
some notions of which the most important are those of elementary
conjunction and elementary disjunction.

The logic product of variables and their negations is referred
to as an elementary conjunction $K = x_1\bar{x}_2 x_3\bar{x}_4 x_5$. For a conjunction
to be elementary, the following condition should be met: each
variable should enter the product only once. The number of
variables in an elementary conjunction is referred to as its rank.
For equivalent transformations of logic functions, notions of
neighboring and isolated conjunctions are often used. Two conjunctions of the same rank and involving the same variables are
termed *neighboring* if they differ only in one inverted variable,
i.e., if one involves a jth variable without the negation sign, and
another has this variable with the sign. For instance, the fourth-
rank conjunctions $K_1^{(4)} = x_1\bar{x}_2 x_3\bar{x}_4$ and $K_2^{(4)} = x_1 x_2\bar{x}_3\bar{x}_4$ are
neighboring conjunctions.

Elementary disjunction is defined similarly as the logic sum of direct or inverted variables, each of them occurring only once. The disjunction $D = x_1 \lor \bar{x}_2 \lor \bar{x}_3 \lor \bar{x}_4$ is elementary.

The major objective of logic function minimization is the derivation of forms involving the smallest number of symbols of binary variables or their negations and those of logic operations; i.e., minimization eliminates unnecessary binary variables.

Binary variables and logic formula terms are called unnecessary if they do not influence the value of the formula under consideration. For example, in the formula

$$x_1 x_2 x_3 \lor \bar{x}_1 x_2 x_3$$

x_1 is unnecessary because true values of this formula depend only on x_2 and x_3. Indeed,

$$x_1 x_2 x_3 \lor \bar{x}_1 x_2 x_3 = x_2 x_3 (x_1 \lor \bar{x}_1) = x_2 x_3 \cdot 1 = x_2 x_3$$

This example demonstrates that minimization may be conveniently done through identification of neighboring conjunctions that fuse into a lower level conjunction.

Let elementary conjunctions of the form $K_\alpha(n) = x_1^\alpha \bar{x}_2^\alpha \ldots$ $x_i^\alpha \ldots \bar{x}_n^\alpha$ be given and their join (i.e., disjunction) be determined. Obviously, the disjunction of two fusing conjunctions results in a reduction in the number of elements and of computed logic expressions. Let $K_{\alpha 0}(n)$ be the reference conjunction; i.e., let all other elementary conjunctions be tested on fusion with binary variables of the conjunction $K_{\alpha 0}(n)$. To find the elementary conjunctions of $K_\alpha(n)$ fusing in the element $x_{i0}^{\alpha 0}$ with the reference one, invert it in all the elements but $x_i^{\alpha 0}$. The resulting conjunction is in parallel, projected by a cylindrical lens onto the light-sensitive side of a digital optically controlled SLM. Onto another side of the SLM an operator field is projected, with each of its rows an elementary conjunction $K_\alpha(n)$. If $x = 1/0$ means presence/absence of light, there is no light-reflecting element in that row of the optically controlled SLM that contains a conjunction fusing in $x_{i0}^{\alpha 0}$. By generating in a rowwise manner the attributes

$$P_\alpha = \bigvee_{i=1}^{n} (x_i^\alpha \land \tilde{x}_i^{\alpha 0}) \quad \text{where} \quad \tilde{x}_i^{\alpha 0} = \begin{cases} \bar{x}_i^{\alpha 0} & \text{for} \quad i \neq i_0 \\ x_i^{\alpha 0} & \text{for} \quad i = i_0 \end{cases}$$

one may find the row with neighboring conjunctions. An optical scheme performing this task may be seen in Fig. 7.1.

If rows are desired that coincide with the reference conjunction, all its elements should be inverted.

In an N X N digital SLM, $M = N^2/r$ different elementary conjunctions may be placed, where r is conjunction rank.

Although electronic computers execute this task by comparing
successively stored conjunctions with reference one $K_{\alpha 0}^{(n)}$,
which requires M cycles for M conjunctions, the optoelectronic
processor compares all the conjunctions in one cycle. Assuming
that the clock cycles in the electronic computer and optoelectronic
processor are equal, determine that the gain in throughput is
equal to the number of compared conjunctions. For N = 128 and
r = 32, M = 512. Clearly, this estimate is the upper bound.

Consider now a case in which M conjunctions are to be tested
on fusing with each element $x_i^{\alpha 0}$ of the reference conjunction
(i = 1, ..., n). Obviously, the optoelectronic version takes n
cycles. If in the electronic version $x_i^{\alpha 0}$ is simultaneously compared with terms x_i^{α} occupying the same positions, M cycles are
necessary to compare all the elements of all the elementary conjunctions, hence, the throughput gain if the optoelectronic processor is M/n. For N = 128 and n = 32, the gain is 16. This is
the lower bound.

The above example demonstrates that, in logic minimization,
the optoelectronic processor provides gain in the throughput 10
to 10^2 as compared with the electronic version.

Similarly, a vector is found in a vector array. Let the vector
$X = \{x_1 \ldots x_i \ldots x_n\}$ and $\hat{A} = \{a_{ij}\}$ matrix m X n be given, and
verify whether there is a definite vector X in the set of vectors
represented by matrix $\hat{A}$. As a preliminary, represent the vectors x_i and a_{ij} in the paraphase code.

The vector search algorithm is built around computation of the
following expression:

$$P_i = \bigvee_{j=1}^{n} a_{ij}\bar{x}_j$$

where a_{ij} are elements of the array $\hat{A}$, $\bar{x}_j$ is the inverted element
of the vector X, n is the number of positions in vector X, P_i is
an attribute with respect to which the existence of vector X is
verified and the address in the array (i.e., row number) is determined.

An optical scheme implementing this operation is shown in Fig.
7.2. As in the above case, a vector array is fetched from the
holographic memory by means of a laser beam accessing the appropriate memory location. The fetched vector array is projected onto the input surface of a digital optically controlled SLM.

Vector X is generated by means of a fast electrically controlled SLM addressed columnwise in parallel according to values of $\bar{x}_j$. Vector $\bar{x}_j$ repeated as many times as there are rows in the array a_{ij} (i.e., m times) is projected over the input surface of the digital optically controlled SLM. Reflection from each cell of the SLM occurs only if there are signals on both input and output surfaces.

A vector address in the file is determined by focusing the results of multiplication $a_{ij}\bar{x}_j$ onto a row of photodiodes, through a cylindrical lens, i.e., by performing the operation $V_{i=1}^n a_{ij}\bar{x}_j$. The absence of light on ith element of this row corresponds to $P_i = 0$ and means that ith row of matrix A contains a vector equal to the desired vector.

7.3 EXTENSION OF THE LIST OF OPERATIONS FOR LOGIC COMPUTATIONS

Computations feature great diversity both in executed tasks and in the algorithms employed. There are problems leading to systems of logic rather than algebraic expressions. Equations in which variables may take only two values (0 or 1) and operations that are defined by an algebra based, generally speaking, on an arbitrary basis of operations are regarded as logic operations. The following might be taken as a basis:

1. Negation (NOT):

x	0	1
$\bar{x}$	1	0

2. Multiplication ($\wedge$)

x_1	0	0	1	1
x_2	0	1	0	1
$x_1 \wedge x_2$	0	0	0	1

3. Mod 2 addition ($\oplus$):

x_1	0	0	1	1
x_2	0	1	0	1
$x_1 \oplus x_2$	0	1	1	0

This algebra resembles the Boolean, with the difference that mod 2 addition is substituted for logic addition (disjunction). Other operations are identical to those of the Boolean algebra.

Both general-purpose and dedicated computers are used to solve logic equations. The latter are superior in throughput to general purpose computers, but their scope of application is narrower.

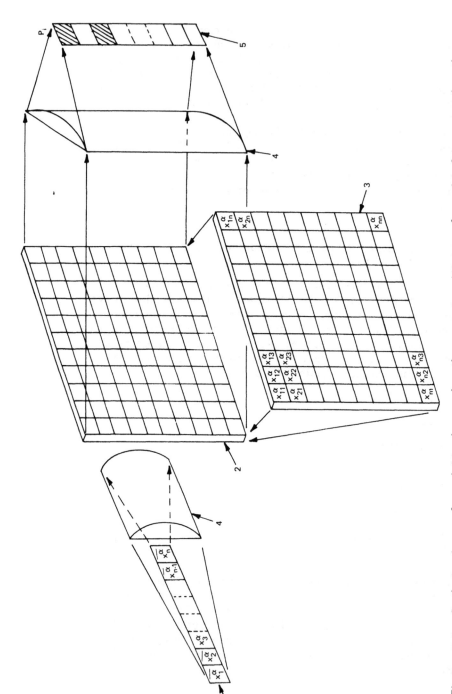

Figure 7.1. Optical scheme of a neighboring conjunction search: (1) transformed reference conjunction (only x_3^α is not inverted), (2) digital optically controlled SLM, (3) operator field with conjunction array, (4) cylindrical lens, and (5) photodiode row. The absence of light in a diode means that conjunctions are slewing.

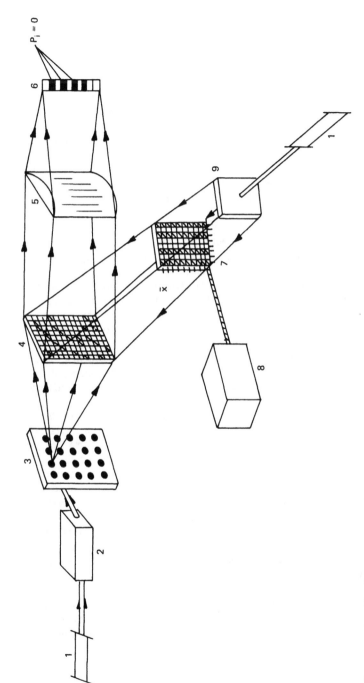

Figure 7.2. Optical scheme of search in vector array: (1) laser, (2) deflector, (3) hologram matrix, (4) digital optically controlled SLM, (5) cylindrical lens, (6) photodiode row, (7) electrically controlled SLM, (8) electronic memory, and (9) beam collimator.

As noted previously, the control-operator optoelectronic processor may handle a wide range of tasks without changing its structure because only control operators are changed in compliance with the current task.

Owing to the peculiarities of logic equations, their solution on the optoelectronic processor may result in appreciable gain as compared with the electronic version. These peculiarities are due to the fact that logic equations enable a high degree of parallelization and that logic operations do not involve carries from position to position.

Since individual positions in logic equations are processed independently, the parallelism inherent to optics can be expected to permit the development of very effective methods for the solution of logic algebra problems.

To make the optoelectronic processor equally effective in dealing with logic and arithmetic problems, one, evidently, must extend the list of elementary logic operations executed by means of control operators. For the above methods of executing basic elementary logic operations, such as multiplication, addition, and mod 2 addition, corresponding optical schemes were presented, some of them making use of the paraphase code and others employing the conventional code. This approach leads to some complications in the optical outlay. Paraphase representation of all the variables involved in computations enables the uniform computation of such functions as $F = V_{i=1}^{n} x_n{}^{\alpha}$ through the control operator method, where $x_n{}^{\alpha}$ may be either x_n or $\bar{x}_n$ and V indicates the logic sum.

Operator fields for computation of all possible functions $F = x_1{}^{\alpha} \vee x_2{}^{\alpha}$ of variables x_1 and x_2 are shown in Fig. 7.3. Discrete picture operators and binary variables in paraphase form are projected, respectively, on control and controlled surfaces of the digital SLM. An operator's form uniquely defines the type of disjunction over variables. The picture reflected by SLM is collected by the cylindrical lens, light in its focus meaning that disjunction is equal to 1; otherwise, it is 0.

It is clear from the picture which operators are required for implementation of the functions $x_1 \vee x_2$, $\bar{x}_1 \vee x_2$, $x_1 \vee \bar{x}_2$, and $\bar{x}_1 \vee \bar{x}_2$. In these examples, as above, the SLM reflects light only if there are both control and reading radiations.

Figure 7.4 shows the block diagram of an optoelectronic computing unit executing an extended set of basic logic operations. Laser radiation is deflected to the desired memory cell hologram. The picture reconstructed from it is control operator projected onto the input surface of the optical digital SLM.

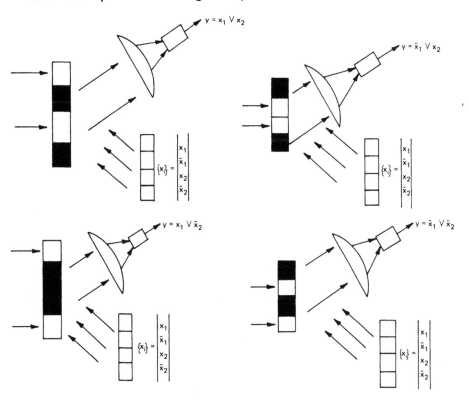

Figure 7.3. Operator fields for all possible functions of the form $y = x_1^\alpha \lor x_2^\alpha$.

Variables to be processed are fetched from the electronic memory to the electrically controlled SLM generating pages of variables. Since variables should be presented in the paraphase code, the first two rows of the electrically controlled SLM are filled by the first row of the first array, the next two rows are filled by the first row of the second array, and the process is repeated for the following rows of processed arrays. Light reflected by the digital optically controlled SLM is collected by a lens matrix and projected either onto the input surface of the SLM of the next processing stage, or onto a photodiode matrix. In the majority of cases, the output of the first SLM should be sent both to the input of the next processing stage and to the photodiode matrix used for, say, correctness check or for error correction.

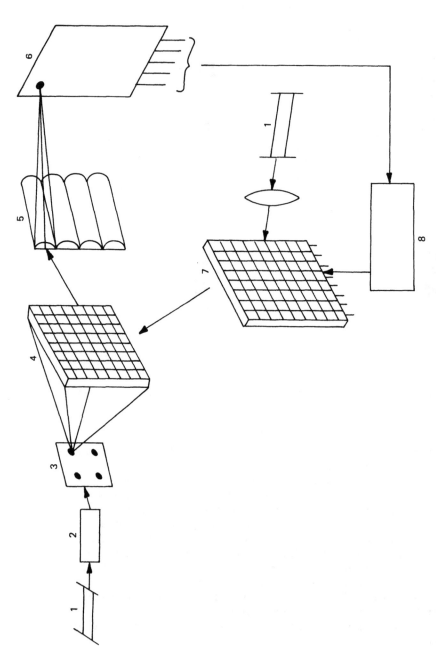

Figure 7.4. Block diagram of an optoelectronic computing unit executing an extended set of basic logical operations: (1) laser, (2) deflector, (3) hologram matrix, (4) optically controlled digital SLM, (5) cylindrical lens matrix, (6) photodiode matrix, (7) electrically controlled SLM, and (8) electronic memory.

The output of photodiodes again is sent to the electronic memory, where it may be stored, accessed by other devices, or again sent to data processing units as indicated by the control unit.

Consider some examples of logic operations over two numerical arrays in an optoelectronic processor with an extended logic basis.

Matrix Negation

Let binary matrix $\hat{M}_1 = \{a_{ij}\}$ be given, and a matrix $\hat{M}$ be desired whose entries are each a negation of the appropriate entries of matrix $\hat{M}_1$, i.e., matrix $\hat{M} = \hat{\bar{M}}_1 = \{\bar{a}_{ij}\}$ is desired. Note that $\bar{a}_{ij} = a_{ij} \vee 0$; hence, to obtain negation, one has to compute disjunction of matrix $\bar{M}_1$ and zero matrix. An operator field for disjunction $\bar{x}_1 \vee x_2$ is shown in Fig. 7.3, and zero matrix should be stored in electronic memory.

Logic Matrix Multiplication

For bit-by-bit multiplication of two matrices $\hat{M}_1 = \{a_{ij}\}$ and $\hat{M}_2 = \{b_{ij}\}$, one should compute matrix $\hat{M}_3$, each of whose entries is a conjunction of the corresponding entries of matrices $\hat{M}_1$ and $\hat{M}_2$; i.e., matrix $\hat{M}_3 = \{c_{ij}\} = \{a_{ij} \wedge b_{ij}\}$ is desired. Logic multiplication $c_{ij} = a_{ij} \wedge b_{ij}$ is representable also as $c_{ij} = \overline{\bar{a}_{ij} \vee \bar{b}_{ij}}$; clearly, c_{ij} may be computed in two clock cycles: in the first cycle, $\bar{a}_{ij} \vee \bar{b}_{ij}$ is computed, and in the second cycle it is negated.

Implication of Two Matrices

Since implication is $a_{ij} \rightarrow b_{ij} = \bar{a}_{ij} \vee b_{ij}$, it may be implemented in a single cycle. An operator field implementing the implication of two elements is shown in Fig. 7.3. The matrix $\hat{M}_3 = \hat{M}_1 \rightarrow \hat{M}_2$ results from projection of a matrix of operators corresponding to the implication of each of the pairs of entries a_{ij} and b_{ij} onto the input surface of an optically controlled SLM.

Equivalence of Two Matrices

The equivalence of two variables is defined as $c_{ij} = a_{ij} \sim b_{ij}$. To define the equivalence of two matrices, let us make use of the fact that $c_{ij} = a_{ij}b_{ij} \vee \bar{a}_{ij}\bar{b}_{ij}$. Denote by $y_{ij} = a_{ij} \vee b_{ij}$ and by $z_{ij} = \bar{a}_{ij} \vee \bar{b}_{ij}$. Then, $c_{ij} = \bar{y}_{ij} \vee \bar{z}_{ij}$. Indeed,

$$\bar{y}_{ij} \vee \bar{z}_{ij} = \overline{(a_{ij} \vee b_{ij})} \vee \overline{(\bar{a}_{ij} \vee \bar{b}_{ij})} = \bar{a}_{ij}\bar{b}_{ij} \vee a_{ij}b_{ij} = c_{ij}$$

Thus, the equivalence of two matrices may be computed in three cycles during which disjunctions y_{ij} and z_{ij} and matrix $\hat{M}_3 = \hat{M}_1 \sim \hat{M}_2$ are determined successively.

Mod 2 Addition of Two Matrices

Mod 2 addition is representable as

$$c_{ij} = a_{ij} \oplus b_{ij} = \bar{a}_{ij}b_{ij} \vee a_{ij}\bar{b}_{ij} = \overline{(a_{ij} \vee \bar{b}_{ij})} \vee \overline{(\bar{a}_{ij} \vee b_{ij})}$$

Thus, if two first cycles are dedicated to the successive computation of the logic sums $y_{ij} = a_{ij} \vee b_{ij}$ and $z_{ij} = \bar{a}_{ij} \vee b_{ij}$, the third cycle generates the matrix $\hat{M}_3 = \hat{M}_1 \oplus \hat{M}_2$, with entries

$$c_{ij} = \bar{y}_{ij} \vee \bar{z}_{ij} = a_{ij} \oplus b_{ij}$$

The above operations over matrices occur inevitably in solutions of logic equations. In the most general form, a polynomial logic nth degree equation has in our algebra the following form:

$$\hat{A}_1{}^0 X \oplus (\hat{A}_2{}^1 X) \wedge (\hat{A}_2{}^2 X) \oplus \cdots \oplus (\hat{A}_n{}^1 X) \wedge \cdots$$

$$\wedge (\hat{A}_n{}^n X) = B \qquad\qquad (7.1)$$

where $\hat{A}_i{}^j$ are matrices of constants, X is an unknown column vector (argument), and B is a constant column vector (free term). The linear matrix equation $\hat{A}X = B$, which is equivalent to the system of linear equations

$$a_{11}x_1 \oplus a_{12}x_2 \oplus \cdots \oplus a_{1j}x_j \oplus \cdots \oplus a_{1n}x_n = b_1$$

$$a_{i1}x_1 \oplus a_{i2}x_2 \oplus \cdots \oplus a_{ij}x_j \oplus \cdots \oplus a_{in}x_n = b_i$$

$$a_{m1}x_1 \oplus a_{m2}x_2 \oplus \cdots \oplus a_{mj}x_j \oplus \cdots \oplus a_{mn}x_n = b_m \qquad (7.2)$$

is the simplest case of Eq. (7.1). It is assumed here that $m = n$; i.e., the lengths of matrix column vectors and row vectors are equal.

If vectors are sufficiently long (10 to 40 bits), the solution of Eq. (7.1) becomes by the fourth order rather a difficult task. Nevertheless, the majority of problems met in electrical circuit troubleshooting, e.g., in automata theory, lead to Eqs. (7.1) and (7.2). Because logic operations over matrices and vectors are executed by the optoelectronic processor in 1 to 3 cycles through the control operator method, it may be expected that the throughput of the optoelectronic processor over logic algebra problems will approach that of fast digital optically controlled SLM, i.e., 10^9 to 10^{10} bps or more if several SLM are used in the processor.

7.4 LOGIC MATRIX BY VECTOR MULTIPLICATION

Logic multiplication of a matrix by a vector is one of the basic operations required for the solution of logic equations. Consider
an algorithm for this operation that allows for a high degree of
parallelism in the optoelectronic processor.

Let the square $\hat{A}$ matrix with n X n entries a_{ij} and a set of
vectors X with n elements x_i each be given. a_{ij} and x_i take the
values 0 and 1, and indices i and j denoting numbers of rows and
columns run from 1 to n.

Logic multiplication of matrix by vector $\hat{A}X = G$ lies in computing the elements g_i of vector G:

$$g_i = \overset{\oplus n}{\underset{j=1}{\Sigma}} a_{ij}x_j \qquad (7.3)$$

where $\overset{\oplus}{\Sigma}$ designates mod 2 addition.

In order to use the throughput of the optoelectronic processor
to full advantage, it is reasonable to execute computations simultaneously over quite a few vectors X. Let there be an array of
vectors X, and execute an operation over each vector. To this
end, take arbitrarily a group of n vectors X and assign the or-
dinal number k to each of them; i.e., denote the vector as X^k
and its element as x_j^k. The results will be denoted as G^k and
g_i^k, respectively.

The multiplication of a matrix by a vector will be performed
simultaneously over n vectors in n clock cycles. In each of n
cycles, the array $\hat{B}_j$ consisting of n similar rows is used as the
input operand, with each row composed of similar positions of n
vectors X^k:

$$\hat{B}_j = \begin{bmatrix} x_j^1 & x_j^2 & \cdots & x_j^k & \cdots & x_j^n \\ \cdots\cdots\cdots\cdots\cdots\cdots\cdots \\ \cdots\cdots\cdots\cdots\cdots\cdots\cdots \\ x_j^1 & x_j^2 & \cdots & x_j^k & \cdots & x_j^n \end{bmatrix} \text{ n similar rows} \qquad (7.4)$$

where index k denotes the number of vector X in the selected
group. The number of such arrays $\hat{B}_j$ for a set of n vectors
X^k is equal to n.

Further, rowwise conjunction of the elements of the jth column of matrix $\hat{A}$ and matrix $\hat{B}_j$ is done:

$$
\hat{C}_j =
\begin{bmatrix}
a_{1j} \wedge (x_j^1 x_j^2 \cdots x_j^k \cdots x_j^n) \\
a_{2j} \wedge (x_j^1 x_j^2 \cdots x_j^k \cdots x_j^n) \\
\cdots\cdots\cdots\cdots\cdots\cdots\cdots \\
a_{nj} \wedge (x_j^1 x_j^2 \cdots x_j^k \cdots x_j^n)
\end{bmatrix}
$$

$$
=
\begin{bmatrix}
a_{1j} \wedge x_j^1 & a_{1j} \wedge x_j^2 & \cdots & a_{1j} \wedge x_j^k & \cdots & a_{1j} x_j^n \\
a_{2j} \wedge x_j^1 & a_{2j} \wedge x_j^2 & \cdots & a_{2j} \wedge x_j^k & \cdots & a_{2j} \wedge x_j^n \\
\cdots\cdots & \cdots\cdots & & \cdots\cdots & & \cdots\cdots \\
a_{nj} \wedge x_j^1 & a_{nj} \wedge x_j^2 & \cdots & a_{nj} \wedge x_j^k & \cdots & a_{nj} \wedge x_j^n
\end{bmatrix}
\quad (7.5)
$$

If there are zero entries in the jth column of matrix $\hat{A}$, a zero row occurs correspondingly in matrix $\hat{C}_j$. This is, essentially, the difference between matrices $\hat{B}_j$ and $\hat{C}_j$.

If one now performs elementwise mod 2 addition of n matrices $\hat{C}_j$ (j = 1, 2, ..., n), i.e., computes a new matrix in n cycles,

$$
\hat{G} = \overset{\oplus n}{\underset{j=1}{\Sigma}} \hat{C}_j
$$

each column of this matrix consists of elements of vector G^k according to

$$
g_i^k = \overset{\oplus n}{\underset{j=1}{\Sigma}} a_{ij} x_j^k
$$

As in a cycle of the optoelectronic processor, arrays $\hat{B}_j$, $\hat{C}_j$ may be generated and matrices $\hat{C}_j$ and $\hat{C}_{j+1}$ may be added, mod 2 and computation of n vectors G^k requires n cycles. Thus, one sees that logic matrix-by-vector multiplication is done effectively in one cycle.

It is readily apparent that, under fixed sizes of matrices and vectors, throughput may be enhanced by increasing the data capacity of the field, i.e., the dimensions of the digital optically controlled SLM. For instance, for the n X n matrix $\hat{A}$ and 2n X 2n SLM, computations may be executed simultaneously over 4n vectors (four groups of n vectors). In doing so, each of the n arrays $\hat{B}_j$ will consist of four subarrays, each containing n similar rows composed of the same positions of n vectors from one of the four groups. Matrix $\hat{C}_j$ results from conjunction of the jth column of the matrix over all four subarrays comprising $\hat{B}_j$. This

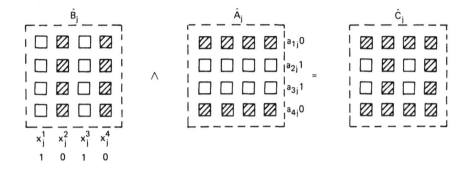

Figure 7.5. Scheme of conjunction of matrices $\hat{A}_j$ and $\hat{B}_j$. 1s and 0s are shown by light and black squares, respectively.

is followed by n cycles of elementwise mod 2 addition of the sequence of n pictures C_j. 4n vectors require n cycles; i.e., computation of one vector is effectively done in one-fourth of the clock cycle. Generally, for matrix and vector dimensionality equal to n and for fixed data field capacity $N^2 = Ln \times Ln$ where L is an integer, the effective part of the cycle for one vector is

$$m = \frac{\text{number of computational cycles in a period}}{\text{number of vectors computed in a period}} = \frac{1}{L^2} \leq 1$$

Multiplication of m^{-1} by f_T results in the throughput P; i.e., the number of operations per second is

$$P = L^2 f_T \quad \text{ops}$$

and the data processing speed is

$$V = L^2 n f_T \quad \text{bps}$$

If the dimensionality of vectors and matrices is n = 32, the controlled SLM field capacity is $N^2 = 128 \times 128$ elements, and the clock frequency is $f_T = 10$ Mcps, the computation throughput is $P \approx 1.6 \times 10^8$ ops and the processing speed $V \approx 5 \times 10^9$ bps. If n = 16, $P \approx 6.4 \times 10^8$ ops and $V \approx 10^{10}$ bps.

Matrix by vector multiplication may be performed in the optoelectronic processor in various ways, depending on hardware and the methods of computing conjunctions and mod 2 additions.

For example, matrix $\hat{C}_j$ may be computed as the conjunction of two matrices $\hat{B}_j$ and $\hat{A}_j$, where matrix $\hat{A}_j$ consists of n repeated columns a_{ij} (i = 1, ..., n) under fixed j. Figure 7.5 schematically depicts input variables—matrices $\hat{B}_j$ and $\hat{A}_j$ and matrix $\hat{C}_j = \hat{B}_j \wedge \hat{A}_j$.

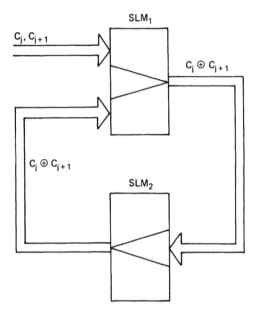

Figure 7.6. Scheme of mod 2 addition of two matrices by means of two digital optically controlled SLM, where SLM 2 operates as optical memory.

The resulting picture of matrix $\hat{G}$ is obtained in n cycles through elementwise mod 2 addition of matrices $\hat{C}_j$. Sequential mod 2 addition of multiple pictures may be done by means of, say, two optically controlled digital SLM, of which one performs desired logic operation over two pictures $\hat{C}_j \oplus \hat{C}_{j+1}$ and the other is used as optical memory, storing partial results of $\hat{C}_j \oplus \hat{C}_{j+1}$. In the next cycle, matrix $\hat{C}_{j+2}$ is added to the stored result. Figure 7.6 shows a diagram of the successive addition of matrices $\hat{C}_j$ by means of a digital optical SLM operating as the optical main memory.

When comparing optoelectronic and electronic matrix by vector multiplications, one should take into account that traditional computer hardware enables parallelization of computations into up to 10^4 paths. However, such a number of paths may be implemented by up to 100 individual devices consisting of thousands of logic gates each. But (1) such an electronic computer is inevitably dedicated; and (2) such a high degree of parallelism seems to be close to the limit, but in the optoelectronic version further improvement of parallelism is possible by using more SLM. The major merit of the optoelectronic control-operator processor is its universality, because its applicability is defined by control operators under a fixed processor structure.

BIBLIOGRAPHY

Meyorov, S. A., and Lee, C. K., On one method of execution of arithmetical and other operations on holographic arrangements, Priborostrolniye *17*, No. 2, p. 58, 1974.

Mnansakanyan, E. A., Morozov, V. N., Popov, V. M., et al., Certain algorithms for getting solutions of problems in optoelectronic processors, DAN USSR *243*, No. 4, 1978.

8

Optoelectronic Processors: Structural and Functional Aspects

8.1 PROCESSOR FUNCTIONAL STRUCTURE

The state of the art and outlooks of computer engineering pose the following major requirements to the optoelectronic control-operator processor:

High degree of parallelism, i.e., the possibility of digital processing of two-dimensional data arrays ranging from 32 X 32 to 128 X 128 discrete binary elements

Array handling speed over 10^{-7} sec

Power consumption at the level of currently designed computer units

Long lifetime

In addition, the design of the optoelectronic processor should preserve its peculiarities, which are at the same time its most important merits, such as the possibility of using multiple (up to 10^3 to 10^4) basic operators, thus providing much higher throughput than electronic computers under the same speed of elements, and the possibility of varying basic operators during computations, thus providing variability of processor functional structure.

With due regard to the above, the functional structure of the optoelectronic processor with control operators may consist of the following units (Fig. 8.1): controlled light beams, hologram records, data carrier, digital optically controlled SLM, and buffer memory.

The controlled light beam unit (CLBU) generating vectored light beams consists of laser, deflector, electronic controller, and an optical scheme usually used for holographic data recording.

The hologram recording unit (HRU) is intended for generation of data pages containing basic operators, program parts, and other numerical files and also for optical recording of this information in

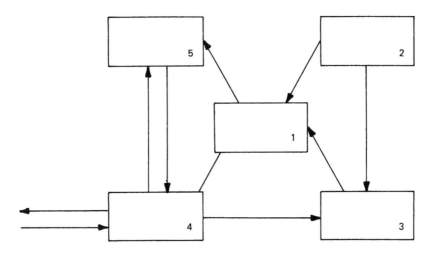

Figure 8.1. Functional diagram of a control-operator optoelectronic processor: (1) data carrier unit, (2) controlled light beam
unit, (3) hologram recording unit, (4) buffer memory unit, and
(5) optically controlled SLM unit.

the form of Fourier holograms in some data-carrying medium. An
HRU consists of data page composer, digital SLM addressed op-
cially or electrically, and an appropriate optical circuit.   The
controlled light beam unit serves for data reading out of SLM and
writing it as holograms into the data carrier unit (DCU), as well
as for hologram cancellation and re-recording. Thus, the totality
of CLBU, HRU, and DCU resembles existing optical holographic
memories with partial data re-recording (PROM) for long-time
storage of operator fields, programs, and data arrays in the form
of two-dimensional pictures.

The buffer memory unit (BMU) stores partial results.

The optically controlled SLM unit (OCSLMU) performing optical processing of numerical pictures consists of several optically
coupled fast-digital optically controlled SLM. The OCSLMU comprises some optical elements, lasers, and optically controlled digital SLM that enable logic manipulations with brightness amplification of data and pictures.

The arrows in Fig. 8.1 indicate functional relations between
these processor units.

Control operators stored in the DCU define the algorithms of
operation execution. Operators are first generated by the HRU's

controlled SLM and then are written as holograms by the CLBU in the DCU areas according to addresses input into the HRU from the environment. These operators may be modified and improved in the course of task execution.

The optoelectronic processor may concurrently operate in three modes: data input, data processing, and data output. In the first mode, input data (e.g., a control-operator data processing program) are written into the DCU through the BMU and the HRU by means of an input program stored in the DCU. In the second mode, digital pictures are processed in the OCSLMU according to appropriate programs, with partial results sent to the BMU. Some results may be written into the DCU if required. For example, new control operators may be generated or existing operators may be modified as part of program execution. Since the operator defines the nature of the executed operation, at the logic level new operations may be generated as part of processing. This new feature of the optoelectronic processor is not known in electronic computers.

In the data output mode, results stored in the DCU and the BMU are sent under program control to external devices.

8.2 DESIGN OF OPTOELECTRONIC PROCESSOR UNITS: A STRUCTURAL APPROACH

The design of the optoelectronic processor and methods of program execution are in the main defined by the elements used, i.e., by the performance of the digital controlled SLM. Decisions about the role of the optoelectronic processor in the execution of a general task are made depending on available technology and the requirements for fast operation.

The control-operator optoelectronic processor has of course the greatest functional possibilities in diverse applications. Because the computational structure of such a processor may vary, it may underlie self-learning and self-organizing computers. But this does not imply that dedicated optoelectronic units will not be designed.

The above indicates that synthesis methods should be developed for optoelectronic devices executing logic operations over digital pictures.

For example, consider the synthesis of an arithmetic adder built around digital optical SLM. To provide a basis for comparison, synthesis will rely upon an algorithm of adding as used in electronic computers.

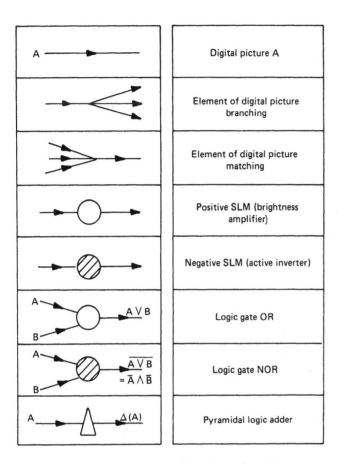

Figure 8.2. Designations of basic logical elements.

Two types of digital optically controlled SLM will be considered: a two-input positive SLM for brightness amplification, and a two-input negative SLM operating as an active inverter. Obviously, two digital pictures presented to the inputs of positive SLM are ORed in each cell, and the negative SLM performs NOR.

In Fig. 8.2, the graphic symbols of basic logic elements are shown along with their functions. In addition to logic manipulations of pictures, optical operations are also required. They are performed when digital pictures are transmitted from one logic element to another and are used actually as a means of optical

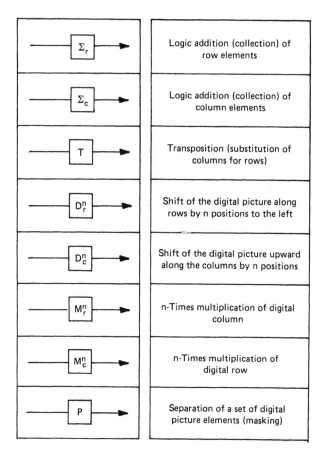

Σ_r	Logic addition (collection) of row elements
Σ_c	Logic addition (collection) of column elements
T	Transposition (substitution of columns for rows)
D_r^n	Shift of the digital picture along rows by n positions to the left
D_c^n	Shift of the digital picture upward along the columns by n positions
M_r^n	n-Times multiplication of digital column
M_c^n	n-Times multiplication of digital row
P	Separation of a set of digital picture elements (masking)

Figure 8.3. Optical transformations and their designations.

path switching. In contrast to logic transformations, optical transformations do not consume computer time. A set of required optical transformations and their symbols may be seen in Fig. 8.3.

Binary numbers are arranged along the rows of the digital picture so that row length is a multiple of n, the number of bits in
one number, and certain columns correspond to the same number
positions. An "extended" digital picture is also considered, where
an n X n "number cell" is allocated to each n-bit number.

All optical manipulations with numbers are performed within
their cells. For instance, the projecting operator P converts

into zero all the positions of a number cell but indicated ones. This operation is done by means of a permanent optical mask repeated in all the number cells. Generally, a q-bit shift along rows $D_r{}^q$ or columns $D_c{}^q$ is performed by shifting bits within fixed boundaries of number cells rather than by shifting the entire picture. Elements set free as the result of shift are filled with zeros; bits going outside cell boundaries disappear without transferring to other cells. The operations $D_r{}^q$ and $D_c{}^q$ may be implemented by mirrors or fiber optics. Multiplication M^p lies in the repetition of a number in the first row in $(p - 1)$ rows of the extended number cell. This may be done, say, by extending the picture over p rows through a cylindrical lens raster. Obviously, rowwise $\Sigma_r{}^q$ or columnwise $\Sigma_c{}^q$ contraction of the extended picture is performed through a reverse lens raster.

The kth position of the arithmetic sum of two numbers $A = a_n a_{n-1} \ldots a_1$ and $B = b_n b_{n-1} \ldots b_1$ is representable as

$$s_k = a_k b_k \ell_k + \bar{a}_k \bar{b}_k \ell_k + \bar{a}_k b_k \bar{\ell}_k + a_k \bar{b}_k \bar{\ell}_k \qquad (8.1)$$

From here on, product, sum, and $\bar{a}$ designate, respectively, conjunction, disjunction, and inversion of the number a.

A carry to the kth position ℓ_k is expressed in terms of own carries c_i and carry tokens t_i in the ith position:

$$c_i = a_i b_i \qquad t_i = a_i + b_i \qquad (8.2)$$

$$\ell_k = c_{k-1} + t_{k-1} c_{k-2} + t_{k-1} t_{k-2} c_{k-3} + \ldots + t_{k-1} \ldots t_1 c_0 \qquad (8.3)$$

where c_0 is a carry added to the lowest position if the inverse code is obtained or the number is rounded off.

Since the following holds:

$$\bar{\ell}_k = \overline{c_{k-1} + t_{k-1} \ell_{k-1}} = \bar{t}_{k-1} + \bar{c}_{k-1} \bar{\ell}_{k-1} \qquad (8.4)$$

one may use

$$xyz = \overline{\bar{x} + \bar{y} + \bar{z}}$$

to obtain logic expressions for sum (8.1) and carries (8.3) in terms of inversions and disjunctions only:

$$s_k = \overline{\bar{a}_k + \bar{b}_k + \bar{\ell}_k} + \overline{a_k + b_k + \bar{\ell}_k} + \overline{\bar{a}_k + b_k + \ell_k} +$$

$$\overline{a_k + \bar{b}_k + \ell_k} \qquad (8.5)$$

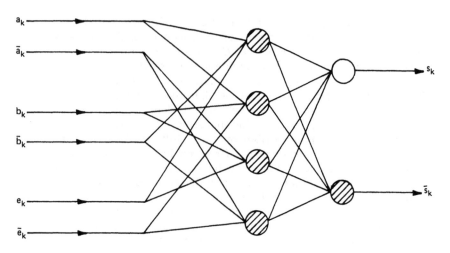

Figure 8.4. Two-cycle optoelectronic three-input digital picture adder in designations of Fig. 8.2.

$$\ell_k = \sum_{i=1}^{k} \overline{\Delta_{k-1,i}(\bar{t}) + \bar{c}_{i-1}} \qquad (8.6)$$

$$\bar{\ell}_k = \sum_{i=1}^{k} \overline{\Delta_{k-1,i}(c) + t_{i-1}} \qquad (8.7)$$

where $c_0 \equiv \ell_1$ and vector components $\Delta_k(x)$ are defined as follows:

$$\Delta_{k,i}(x) = \begin{bmatrix} \Delta_{k,k}(x) \\ \Delta_{k,k-1}(x) \\ \cdots \cdots \cdots \\ \Delta_{k,1}(x) \end{bmatrix} = \begin{bmatrix} x_k \\ x_k + x_{k-1} \\ \cdots \cdots \cdots \cdots \\ x_k + x_{k-1} + \cdots + x_1 \end{bmatrix} \quad (i \leq k) \ (8.8)$$

Figure 8.4 demonstrates in terms of Figs. 8.2 and 8.3 a scheme of a two-cycle, three-input adder of digital pictures. A scheme for simultaneous generation of carries is shown in Fig. 8.5 along with a sequence of transformations of the content of a four-bit number cell. The projection P_Δ used in this scheme converts into zero all the picture elements lying below the main diagonal.

Carries required for the computation of components of the vector $\Delta_{k,i}(x)$ may be generated by a so-called pyramidal adder Δ. A scheme of operator Δ depicted in Fig. 8.6 consists of a positive

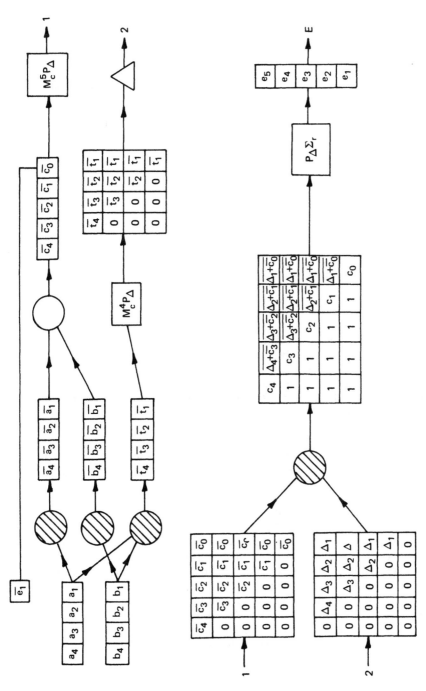

Figure 8.5.   Optoelectronic processor with simultaneous carry generation.

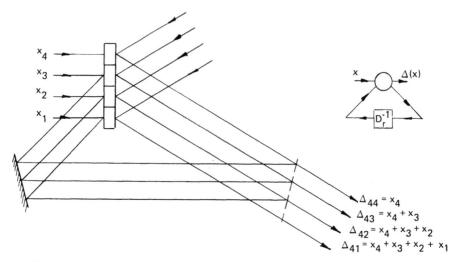

Figure 8.6. Optical computation of components of vector $\Delta_{k,i}(x)$.

SLM and mirrors providing feedback with a one-position right shift. It may be readily seen that operator Δ repeats 1 in all the lower positions of number $\Delta_k(x)$ beginning from that position x where it occurs for the first time. As the time of light propagation through all the positions of number $\Delta_k(x)$ in the optical feedback is essentially smaller than the SLM switching time, 1s are generated in all the positions in essence simultaneously, i.e., in a single clock cycle.

The combination of an adder (Fig. 8.4) and a carry-generating circuit (Fig. 8.5) results in a parallel adder generating sum in six clock cycles, four cycles for carries and two for the sum.

In computer engineering there are many arithmetic adders; e.g., circuits with simultaneous generation of carries for byte groups and superparallel adders. The selection of a specific adder depends on the designer's objectives, whether minimal hardware (e.g., total number of SLM cells) or time reduction is required. Adder design depends also on the particular functionally complete set of logical elements used.

The above adder for two 32-bit numbers requires about 3400 SLM cells. Although a similar electronic adder requires about 400 transistor logic elements, like OR and AND, it requires about 1000 individual inputs to logic elements; digital SLM are uniformly structured and have four picture inputs at most. Owing to a high degree of parallelism, the total throughput of an optical adder may be sufficiently high.

This example demonstrates that SLM-based optical computations are effective for parallelization of arithmetic operations over multidigit numbers even if they rely on traditional electronic algorithms for the execution of these operations. This is possible because an appreciable portion of computational operations may be reduced to optical operations over digital pictures, such as multiplication, contraction, shift, and masking, which do not need computer time and logic elements.

Optoelectronic arithmetic addition in a control-operator processor is likely to be more effective because it may also perform numerous other logic and arithmetic operations. As for the optoelectronic device dedicated to arithmetic addition only, its effectiveness may be significantly improved by using other number systems without long carries such as residue number system.

8.3 INTERCONNECTIONS BETWEEN THE OPTOELECTRONIC PROCESSOR AND ELECTRONIC COMPUTERS

The optoelectronic processing of digital pictures enables an important step in the development of digital computer structure. Current electronic computers are intended mostly for executing operations over binary multidigit numbers. Meaningful data relations between digits are hard wired. The development of a computer operating over two-dimensional digital pictures would permit one to take into account more profound data relations than those allowed for in existing computers. The relations between numbers representing values for a function might be taken into consideration, as well as relations between individual positions of a number.

There are read-only holographic memories (ROM) in the optoelectronic processor that store a set of control operators. The operators are read from ROM and projected onto the control surface of an optically controlled SLM, thus defining the type of operation executed in the optical data processing path. The number of executed operations defined by the memory capacity may be rather large. It should also be noted that the optical path structure may be defined by the class of applications.

The optoelectronic processor has the highest throughput in the case of a rather uniform processing of large data arrays. It is difficult and unreasonable to use it for complicated and diverse manipulations with small amounts of data, for example, to process program instructions. Modern electronic computers are best suited to this task.

In this connection, it is expedient to use the optoelectronic processor for uniform processing of large data arrays and to leave to its electronic counterpart control and result handling operations, i.e., to utterly separate data and instruction flows.

A present, the technology of holographic ROM is rather evolved, and media for fast read/write memories are nonexistent. Therefore, it is expedient to employ in optoelectronic processors a two-level structure of semiconductor large-format memories. Electronic memories are coupled to the optical data processing path through interfaces that convert electrical signals into optical signals, and vice versa, and transform their representation from one- to two-dimensional and the other way round. Signal transformations are performed by electrically controlled digital SLM and photodiode matrices.

Current difficulties may be overcome by structural methods that rely upon computer control of the optoelectronic processor and semiconductor main memory.

Figure 8.7 shows the block diagram of a hybrid optoelectronic processor involving an optoelectronic path for processing large data arrays and a general-purpose computer.

In order that the throughput of the optoelectronic processor not be limited by input/output (I/O) devices (by transformation of electrical signals into two-dimensional digital pictures, and vice versa, in our case), the block diagram has a digital picture switch (DPS) and brancher (DPB).

A digital picture switch is, essentially, an OR gate operating over digital pictures generated by the electrooptical transformer. A digital picture brancher performs a similar function for optoelectrical transformations. Both of them may be implemented by means of, say, fiber optics.

An optical data processing path for a hybrid optoelectronic control-operator processor is shown in Fig. 8.8. It comprises a holographic ROM storing digital pictures of control operators, a digital optically controlled SLM, and a multichannel switch of digital pictures. Logical and arithmetic processing are performed in the unit of the optically controlled SLM (OCSLMU), which matches source digital pictures with those of control operators. The multichannel switch performs logical AND and OR over elements of the resulting digital picture.

The number of optically controlled SLM and holographic memory units depends on the class of logic operations to be performed

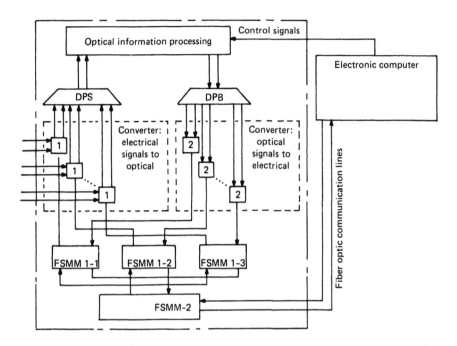

Figure 8.7. Structure of hybrid optoelectronic processor: DPS, digital picture switch; DPB, digital picture brancher; FSMM, fast semiconductor main memory. (1) Electrically controlled SLM, and (2) photodiode matrix.

by the optoelectronic processor. For example, one SLM suffices for logic functions, but at least three SLM are required for arithmetic operations. In Fig. 8.8, SLM 1 amplifies digital pictures, SLM 5 amplifies output optical signals, and processing is properly done by SLM 2, 3, and 4.

Throughputs of the optically controlled SLM unit and hybrid optoelectronic processor may be estimated as follows. OCSLMU processes data in time

$$T_0 = kT_t + T_\ell \tag{8.9}$$

where T_t is the SLM switching time, k is the number of successive SLM required for a given group of operations, and T_ℓ is the light propagation time in the OCSLMU. In the example of Fig. 8.8, $1 \leq k \leq 5$.

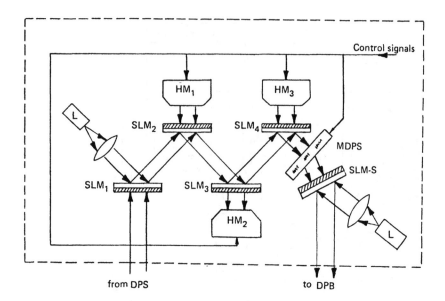

Figure 8.8. Optical data processing path of a hybrid optoelectronic control-operator processor: HM, holographic memory; L, laser; SLM, digital optically controlled space light modulator; DPS, digital picture switch; DPB, digital picture brancher; and MDPS, multichannel DPS.

In time T_0 OCSLMU may process

$$V = \frac{m^2}{k_p} \quad \text{bits} \tag{8.10}$$

where m is the length of an SLM row or column, and k_p is factor allowing for the redundancy of data representation of processing by the control operator technique ($k_p > 1$). OCSLMU throughput is

$$P_0 = \frac{V}{T_0} = \frac{m^2}{k_p T_0} \quad \text{bps} \tag{8.11}$$

The throughput P of an optoelectronic processor with L optical paths processing in parallel n-bit numbers is each estimated as

$$P = \frac{LP_0}{2n} = \frac{Ln}{2k_p T_0} = \frac{Lm^2}{2nk_p T_0} \quad \text{ops} \tag{8.12}$$

where $n = m^2/n$ bits. For the sake of comparison, this expression was derived with allowance for the fact that, in electronic computers, two operands are processed concurrently.

To support the required throughput P, the optoelectronic processor requires $N_{\ell 0}$ electrooptic signal transformers:

$$N_{\ell 0} = \frac{LP_0}{P_{\ell 0}} \qquad (8.13)$$

where $P_{\ell 0}$ is the transformer throughput.

The number of optoelectronic transformers is

$$N_{0\ell} = \frac{LP_0}{P_{0\ell}} \qquad (8.14)$$

where $P_{0\ell}$ is the transformer throughput.

Evidently, $P_{0\ell} = K_c P_{\ell 0}$, where K_c is a factor allowing for data reduction after processing in an OCSLMU ($0 < K_c \leq 1$). OCSLMU throughput P_0, on the one hand, is defined by the requirements for optoelectronic processor throughput P, and, on the other hand, is limited by the characteristics of SLM, photodiode matrices, etc. However, if one takes into consideration the memory capability of digital SLM, it should be readily apparent that the OCSLMU may operate as a pipeline. In this operational mode, the OCSLMU would generate new results at intervals equal to the response time of one SLM, although the total time of result computation would depend on the number of SLM involved. Thus, in the stable mode, OCSLMU throughput is defined only by the value for T_t. For a 1-m optical path in an OCSLMU, $T_\ell \approx 0.3 \times 10^{-8}$ sec. If the SLM switching frequency is 10 Mcps, $T_0 = 10^{-7}$ sec. If arithmetic operations are executed through the control operator method in OCSLMU with at least three SLM, the maximal $k_p = 17$. By assuming for $m^2 = 128 \times 128$ SLM that the mean value of $k_p = 10$ for a considerable body of operations, obtain from (8.11) that

$$P_0 \sim 1.6 \times 10^{10} \quad \text{bps}$$

For $L = 1$ and $n = 32$, the throughput of an optical data processing path as defined by (8.12) is

$$P = 2.5 \times 10^8 \text{ ops} \quad \text{(operations per second)}$$

The throughput P of one path varies from $P_1 = 2.5 \times 10^9$ ops for, say, logic tasks in which input data redundancy is minimal, to $P_2 = 1.5 \times 10^8$ ops for tasks featuring maximal input data redundancy.

The total optoelectronic processor throughput grows with the number of optical data processing paths and the number of elements in a controlled SLM. The number of optical data processing paths is defined by the input data flow, the operations required by processing, the desired processor throughput, and the number of simultaneously processed objects comprising input data array. Moreover, the number of optical paths depends on the possibilities of controlling them and on data exchange with an electronic computer operating together with the processor. Controlled SLM of 10^5 to 10^6 elements may be constructed of modules with, for example, 32 X 32 elements. Thus, a throughput is feasible of over 10^{10} ops.

If the number of optical paths is large enough (e.g., $L > 4$), their totality may be regarded as a multiprocessor optoelectronic common-control computer system featuring

Universality in combination with high effectiveness over various classes of tasks

Feasibility of a high degree of parallelism in data handling

Capability of generating new types of operations at the logical level in the course of computations

Capability of varying the structure of computations in the course of operation

Inherent suitability to both matrix and pipeline processing

BIBLIOGRAPHY

Basov, N. G., Popov, U. M., Morozov, V. N., et al., Principles of construction of optical processors with variable operators, Kvantovaya Electronika 3, p. 526, 1978.

Kartsev, M. A., and Marshalko, B. G., Some questions on the structural organization of special coded optoelectric computer complex, Autometriya 2, p. 3, 1980.

Orlov, L. A., and Svidzenski, K. K., Optoelectronic sum of higher efficiency, Autometriya 3, p. 54, 1978.

Glossary

Algorithm: A prescribed set of precise rules or procedures for problem solution in a finite number of steps.

Antireflection Coating: A special thin optical material layer (single or multilayer) applied to surfaces to reduce the incoming and outgoing optical reflections.

Arithmetic and Logic Unit (ALU): A functional unit of computer system with circuits for the execution of arithmetic and logic operations.

Array Processor: A computer system with numerous ALU (sometimes referred to as elementary processors) concurrently executing operations over data matrix elements.

Associative: Usually refers to memory accessed by its content rather than by location address.

Attenuation: The degree of optical power loss along an optical fiber, fiber bundle, or waveguide, which is the sum of all light scattering and absorption losses, usually expressed in units of decibels per kilometer (dB/km).

Avalanche Photodiode (APD): A photodiode designed to take advantage of avalanche multiplication of photocurrent. As the reverse bias voltage approaches the breakdown voltage, hole-electron pairs created by absorbed photons acquire sufficient energy to create additional hole-electron pairs when they collide with substrate atoms, thereby achieving a multiplication effect.

BER: Bit error rate.

Blazed Grating: A special molded or etched diffraction grating with finely ruled, narrowly spaced, equidistant grooves.

Blocking: The lack of any interconnection between at least two idle lines connected to a network because all possible paths between them are already in use in a manner not permitting interconnection.

Busy Hour: The continuous hour-long period with the maximum average traffic intensity.

Busy Hour Call (BHC): A call during the busy hour used by Europeans as a unit of call intensity.

Call: A discrete engagement of a traffic path.

Call Concentration: The average ratio of calls during BHC to those during the day.

Call-Hour: One or more calls with an aggregate duration of 1 h.

Call Intensity: The number of calls in a group of traffic channels or paths per unit time.

Call-Minute: The quantity represented by one or more calls of aggregate duration of 1 min.

Call-Second: The quantity represented by one or more calls of aggregate duration of 1 sec.

Calling Rate: The call intensity per traffic channel or path during the busy hour.

CCS: The quantity represented by one 100-sec call or by an aggregate of 100 call seconds.

Cladding: The low refractive index material that surrounds the core of the fiber. In plastic-clad silica fibers, the plastic cladding may serve as the coating.

Concentration: In a switching network, the function associated with the network with fewer outlet than inlet terminals.

Control Unit: Controls the operation sequence in a computer and generates the signals required for the execution of operations.

Coordinate Switch: A regular geometric array of crosspoints. For example, a rectangular array, in which, in the case of rectangular one side of the crosspoints, is multiplied in rows and the other side in columns.

Core: The light-conducting portion of the fiber, defined by the high refractive index region. The core is normally in the center of the fiber, bounded by the cladding material.

Crosspoint: A switching device with two states, where one is a low-transmission impedance and the other is a very high one.

Dark Current, I(A): The current that flows in photosensitive detectors when there is no incident radiant flux (total darkness).

Decibel (dB): The ratio of the output optical power (P_O) to the input power (P_i) expressed by 10 log (P_O/P_i).

Detectivity, $D(W^{-1})$: Reciprocal of noise equivalent power: $D = 1/NEP$.

Differential Quantum Efficiency: Used in describing quantum efficiency in devices with the nonlinear output/input characteristics. The slope of the characteristic curve is the differential quantum efficiency.

Digital-Controlled Space Light Modulator: A multielement space-time light modulator whose cells either pass (reflect) light or not, depending on the control signal.

Diode Laser: See Semiconductor Laser.

Dispersion: Used to describe the relationship between refractive index and frequency (or wavelength); a cause of bandwidth limitation in a fiber. Because dispersion causes a broadening of input pulses along the length of the fiber, this mechanism is usually referred to as pulse-spreading. The two major types of dispersion are mode and material.

Dispersion Limited Operation: Used to denote an operation when the dispersion of the pulse, rather than its amplitude, limits the distance between repeaters. In this regime of operation, waveguide and material dispersion are sufficient to preclude an intelligent decision on the presence or absence of a pulse. Correction methods exist by introducing a material in which the dispersion acts in a reverse manner, thereby extending the distance between repeaters.

D-Star, D* ($W^{-1}m^{1/2}Hz^{1/2}$): Detectivity multiplied by the square root of the detector area and the square root of the detector bandwidth.

Duplex: A type of data bus configuration in which terminal devices can receive and transmit at the same time.

Duty Factor, Pulse: The ratio of average pulse duration to average pulse spacing (dimensionless).

Electrooptic: Strictly speaking, this refers to an operation relying on modification of a material's refractive index by electrical fields. In a Kerr cell, the index change is proportional to the square of the electrical field and the material is usually a liquid. In a Pockel's cell, the material is a crystal whose index change is linear with the electrical field. The current engineering usage has become looser and includes optical transmission or reception modified by an electronically controlled device.

Equated Busy-Hour Call (EBHC): One EBHC is the average intensity in one or more traffic paths occupied in the busy hour by a call or calls of aggregate duration of 2 min. Thus, 1E = 30EBHC.

Erlang (E): The dimensionless unit of traffic intensity. It is the intensity in a traffic path such that 1E is the intensity in a path continuously occupied, or in one or more paths carrying an aggregate of a call-hour per hour or a call-minute per minute.

Expansion: In a switching network, the function associated with having more outlet than inlet terminals.

FDM: Frequency division multiplex. In optical communications, one also encounters wavelength division multiplex (WDM). WDM involves the use of several distinct optical sources (lasers), each with a distinct center frequency. FDM may be used with any or all of these distinct sources.

Folded Network: A network in which each trunk or line is connected to both an outlet terminal and an inlet terminal, and which has a capability of completing a path between any two lines, with either terminal of one, to either terminal of the other.

Full-Availability Group: A group of traffic-carrying circuits or trunks in which each circuit or trunk is accessible to all traffic.

Full Availability (Switch) or Fully Switched: A switch or switching network capable of providing a path from every inlet terminal to every outlet terminal (in the absence of traffic considerations).

Functional Units: Basic computer units, such as arithmetic unit, control unit, and memory.

Graded Index: Usually refers to the fiber type wherein the core refractive index decreases almost parabolically radially outward toward the cladding. This type of fiber has the property of combining high-bandwidth capacity with moderately high coupling efficiency.

Grade of Service: A measure of probability that during a peak traffic interval a call will find an idle circuit at the first try.

Holding Time: The duration that a call occupies a traffic path, or the average duration that a call or calls occupy a traffic path or paths.

Holographic Memory: Optical memory storing two-dimensional data in the form of a Fourier hologram.

ILD, Injection Laser Diode: See Semiconductor Laser.

Infinite Sources: A reference to a large number of sources that offer traffic, in comparison to the number of circuits. Numbers of 10 or larger are considered "infinite" ratio for this purpose.

Insertion Loss: The ratio of the output optical power to the input power through an optical coupler device, connector, or splice.

Junction Diode: A special type of semiconductor diode in which the conducting current essentially flows in one direction only.

LED: Light-emitting diode.

Light: For our purposes, radiant energy of wavelengths from about 0.3 to 30 μm; this includes visible wavelengths and those wavelengths, such as ultraviolet and infrared, that can be handled by optical techniques. In more restricted usage, it is the radiant energy within the limits of the visible spectrum.

Light-Emitting Diode: A semiconductor device that emits incoherent light formed at the p-n junction, either from the junction stripe edge or its surface.

Limited-Access Group: A traffic-carrying group of trunks or circuits in which only a fraction is accessible to any traffic source group.

Limited Sources: A traffic-carrying group of trunks or circuit sources not exceeding a selected number.

Linewidth: The frequency range over which most of the energy of the laser beam is distributed.

Lost Calls Cleared: Calls not satisfied at a first attempt may be cleared from the system in current practice and do not reappear during the period of time used for consideration.

Lost Calls Delayed: This assumes that unsatisfied calls are held in the system until satisfied, if they are not done so immediately.

Lost Calls Held: Calls not satisfied at a first attempt are held for a period that, in current practice, does not exceed the average holding of all calls and thus are cleared from the system.

Material Dispersion: Light impulse broadening due to differential delay of various wavelengths of light in a waveguide material. This group delay is aggravated by broad-linewidth light sources.

Modal Dispersion: That component of pulse spreading caused by differential optical path lengths in a multimode fiber.

Modes: Allowed solutions to a mathematical expression describing the propagation of light through a fiber. See Ray for a comparison.

Multimode Fiber: A fiber waveguide that allows more than one mode to propagate.

Noise Equivalent Power (NEP; W/Hz): The root-mean-square (RMS) value of optical power required to produce a unity RMS signal-to-noise ratio.

Nonblocking Network: A network in which at least one available path exists between any pair of idle lines or trunks to be connected.

Nonfolded Network: A network in which each trunk or line is connected to a terminal inlet and/or outlet, and is capable of completing a traffic path between any two lines from the terminal inlet to the outlet of the other terminal.

Numerical Aperture (NA): Measure of light acceptance of a fiber, defined as NA = $(N_1^2 - N_2^2)^{1/2}$, where n_1 and n_2 are, respectively, the refractive index of the core and the cladding. When skew rays are included, the numerical aperture increases.

Occupancy: The traffic intensity in one or more traffic paths. If, for example, there is total occupancy of a traffic path, this is equivalent to 1E.

Optoelectronic Device: A device responsive to electromagnetic radiation (light) in the visible, infrared, or ultraviolet spectral regions; emits or modifies noncoherent or coherent electromagnetic radiation in these same regions; or utilizes such electromagnetic radiation for its internal operation.

Optical Data Processing System: A system in which data are carried by coherent or noncoherent light.

PAMA, Pulse Address Multiple Access: Carriers are distinguished by their temporal and spatial characteristics simultaneously.

Parabolic Profile: Referring to the condition of having the fiber index of refraction vary in a parabolic fashion: $n(r) = n_1$ $[1 - \Delta (r/a)^2]$, where r is the radial distance from the fiber axis and Δ and a are constants; n_1 is the refractive index at r = 0.

Parallel Execution of Tasks: Parallel execution of more than one program by simultaneously operating processors.

Photoconductive Effect: Some nonmetallic materials exhibit a marked increase in electrical conductivity upon absorption of photon energy; this is called the photoconductive effect. The conductivity increase is due to the additional free carriers generated when photon energies are absorbed in electronic transitions. The rate at which free carriers are generated, and the length of time they persist in conducting states (their lifetime), determines the amount of conductivity change.

Photoelectric Effect: Originally referred to all changes in material electrical characteristics due to photon absorption. More recently, it is used to describe the emission of electrons as the result of the absorption of photons in material. This definition is quite broad, since the photons can be of any energy and the electrons can be released into a vacuum or into a second material. The material itself may be a solid, liquid, or gas. With this broad definition, photoconductive, photoelectromagnetic, photoemissive, and photovoltaic effects are all photoelectric.

Photoelectromagnetic Effect: The production of a potential difference by virtue of the interaction of a magnetic field with a photoconductive material subjected to incident radiation.

Photovoltaic Effect: The production of an electromagnetic force (voltage) across a semiconductor p-n junction due to the absorption of photon energy. The potential is caused by the diffusion of hole-electron pairs and hence the phenomenon leads to direct conversion of a part of the absorbed energy into usable electrical force.

Pipelining: Decomposition of arithmetic or logic operation into separate steps executed by the modules of a processor.

Polarization: Used to describe the orientation (in space) of a time-varying electrical or magnetic field vector (including the field vector of an optical signal).

Power, Average: In a pulse laser, the energy per pulse (joules) times the pulse repetition rate (hertz). Expressed in watts.

Power Density: Power per unit area (watts per square meter).

Power, Peak: In a pulsed laser, the maximum power emitted.

Processor: A functional computer unit executing programmed operations and consisting of arithmetic and control units.

Pulse Spreading: The increase in pulse width in a given length of fiber due to the cumulative effect of material dispersion and modal dispersion.

Quantum Efficiency (Dimensionless): A measure of the efficiency of conversion or utilization of optical energy, the number of events produced for each incident quantum.

Quantum-Limited Operation: An operation in which the minimum detectable signal is limited by fluctuations in the average signal current itself.

Radiance: Radiant power per unit source area per unit solid angle, expressed in watts per square meter per steradian.

Radiant Energy: Energy (joules) transferred via electromagnetic waves; with no associated transfer of matter.

Radiant Excitance (W/m^2): Radiant power emitted into a full sphere (4 sr) by a unit area of source.

Radiant Flux: The time rate of flow of radiant energy (watts).

Radiant Intensity (W/sr): The time rate of transfer of radiant energy per unit solid angle.

Ray: A geometric representation of a light path through an optical device; a line normal to the wave front indicating the direction of radiant energy flow. See Modes for comparison.

Rayleigh Scattering: The intrinsic limit of fiber attenuation caused by microcompositional and density nonuniformities; inversely proportional to the optical wavelength to the fourth power.

Refractive Index, n (Dimensionless): The ratio of the velocity of light in vacuum to the velocity of light in the specified medium.

Responsivity (A/W or V/W; sometimes call sensitivity):   The
    ratio of the RMS value of the output current or voltage
    of the RMS value of the incident monochromatic optical
    power.

SDM: Space-division multiplex.

Semiconductor Laser: A laser in which lasing occurs at the junc-
    tion of n-type and p-type semiconductor materials.

Signal-to-Noise Ratio: The optical or electrical power ratio of the
    desired transmitted signal to the undesired noise and cross
    talk.

Single-Linkage Array: The mesh or interconnection spread be-
    tween switching network stages in which every switch is a
    connection to every adjacent stage switch.

Single-Mode Fiber: A fiber waveguide on which only one mode
    propagates.

Skew Rays: Rays skew to the fiber axis. If the fiber waveguide
    is straight, a skew ray traverses a helical path along the fiber,
    not crossing the fiber axis. A skew ray is not confined to the
    meridian plane.

SLD: Superluminescent diode.

SNR: Signal-to-noise ratio.

Space-Division Switching Network: A switching network with
    physically distinct transmission paths.

Spectral Bandwidth (Between Half-Power Points): The wave-
    length interval in which a radiated spectral quantity is not
    less than half its maximum value.

Spectral Irradiance: Irradiance per unit wavelength interval.
    It is measured in units of watts per square meter per mi-
    crometer.

Spectral Radiance (W/sr m^2 μm): Radiance per unit wavelength
    interval.

Step Index Fiber: A fiber in which the core is of a uniform re-
    fractive index.

Step Index Profile: Describing conditions when the refractive
    index changes abruptly from the value n_1 to n_2 at the core-
    cladding interface.

Steradian (sr): The unit solid angular measure, the subtended
    surface area of a sphere divided by the square of a sphere
    divided by the square of the sphere radius. There are 4π
    steradians in a sphere. The solid angle subtended by a cone
    of half-angle θ is 2π (1 - cos θ) steradians.

Switching Network: That part of a switching system that pro-
    vides transmission paths between terminals.

TDM: Time-division multiplex.

Thermal Noise Limited Operation: An operation in which the minimum detectable signal is limited by the thermal noise of the detector and load resistance, and by amplifier noise.

Total Internal Reflection: The reflection that occurs inside an optical fiber in which the incidence angle of the radiation is larger than the critical angle.

Traffic Concentration: The average ratio of the busy hour to the traffic during the day.

Traffic Intensity (Traffic Density): The quantity of traffic in one or more traffic paths in a given unit of time.

Traffic Path: A path over which individual communications pass in sequence, such as in a channel, time slot, circuit, line, trunk, or switch.

Traffic Quantity: The aggregate occupancy time of one or more paths of traffic.

Traffic Rate: The traffic intensity during the busy hour per traffic path.

Traffic Unit: The average intensity (ITU) in one or more traffic paths carrying one call-hour of aggregate traffic. This is usually the busy hour unless specifically stated otherwise.

Unit Call: A unit of traffic intensity whereby one such call is the average intensity in one or more traffic paths occupied during 1 hr by an aggregate of 100 sec.

Vector Processor: A computer system in which multiple ALU are arranged to enable concurrent processing of vector components.

Waveguide Dispersion: That part of the total dispersion attributable to the fact that the critical dimensions of the waveguide, in wavelengths, are a function of frequency.

WDM: Wavelength division multiplex. See FDM.

Index